HOW TO BECOME HAPPILY EMPLOYED

in St. Louis

Barbara Block   and   Janice Benjamin

A STEP-BY-STEP GUIDE TO FINDING THE JOB THAT IS RIGHT FOR YOU

How To Become
Happily Employed
In St. Louis

Barbara Block and Janice Benjamin

Fourth Edition

Copyright 1987, 1986, 1984
by Barbara Block and Janice Benjamin

Library of Congress 84-070522
ISBN 0-9613630-3-7

Cover Design: Steve Hermes

Printed in the United States of America
by Interstate Book Manufacturers, Inc.

Career Management Press
Kansas City, Missouri

ABOUT THE AUTHORS

BARBARA BLOCK

Barbara Block is a business journalist, consultant and speaker whose skills in Career Development and Human Resource Training are sought by a wide range of individuals and organizations. As founder and past president of Kansas City's Career Management Center, she created the first full service organization in that city in 1980, specializing in career planning and job search programs.

In 1982, Ms. Block relocated to San Francisco in order to devote her time to writing, publishing and lecturing. In addition to consulting, she is currently working on a book about Career Transition.

Ms. Block has appeared on numerous radio and television programs, and currently writes a career column for the San Francisco Business Times. She travels around the country to speak, consult, and conduct seminars for business, government, colleges, and professional associations. She holds a M.A. in counseling psychology.

JANICE Y. BENJAMIN

As President and owner of Kansas City's Career Management Center, Janice Y. Benjamin is a practicing career counselor and business consultant. She helps individuals find satisfying career directions, assists employers in stimulating productivity, and serves organization's outplacement needs.

Much in demand as a consultant to businesses and organizations, Ms. Benjamin designs and conducts numerous seminars and workshops. She also speaks to professional associations, government agencies, and educational institutions.

Ms. Benjamin has served as a resource for newspaper articles and has appeared on television and radio programs. She was selected as an Outstanding Young Woman of America for 1983.

A graduate of Tulane University, Ms. Benjamin spent seven years in secondary education. She holds a M.A. in Guidance and Counseling from the University of Missouri-Kansas City.

Other Editions of

**HOW TO BECOME
HAPPILY EMPLOYED**

National Edition

**with San Francisco
Job Hunter's Resource Guide**

**with Kansas City
Job Hunter's Resource Guide**

We gratefuly acknowledge Yehudah Vinitsky and Bert Benjamin for their support and patience; Steve Hermes for his special talent in design and layout; Evie Schmidt and Carol Burrington for their loyal administrative assistance; Corporate Graphics for Typography; Shelly Shrock for Art Production; Mark Klugman for his research; and Linda Goldberg, for her ideas and suggestions; and especially to all our clients, who have taught us so much.

Contents

INTRODUCTION

Looking for work can be pretty grueling. We understand, because for the past 10 years we have helped people figure out what they want to do and then taught them how to find a job and establish a career plan.

We have learned a lot in those years. That's why we wrote this book. We want to show you that by following a methodical step-by-step procedure, you can minimize the fear and confusion while making the task of job hunting not only manageable but even enjoyable.

We are firmly convinced that with the right preparation anyone who wants to can become happily employed.

"Just a minute!" you exclaim, remembering the hopelessly discouraging unemployment statistics you read in the morning paper. "I am not sure I can even find a job, let alone a good one, I'll just stay where I am or take what I can get."

Wait! Before you close this book, let's look at a few facts:

FACT: Over one hundred million Americans are working.

FACT: Six to eight percent unemployment also means at least ninety-two percent of the population is with a job.

Now, consider how most people find jobs:
- Six percent of jobs are found through private employment agencies.
- Twenty-two percent through public agencies.
- Thirty-four percent through ads.
- Seventeen percent through friends and relatives.
- Seventy-seven percent through employers directly.
- Four percent through other methods such as trade union hiring.

What does this tell you?

FACT: Over seventy-seven percent of all jobs are found through *direct contact* with friends, relatives or the person doing the hiring.

What this adds up to is:
- There *are* jobs.
- The majority of openings are discovered through an individual's own networking efforts.

---

* *Source for above statistics: The U.S. Department of Labor, Bureau of Labor Statistics.*

Note: *The percent for jobseekers using each jobseeking method will always total more than 100% because many jobseekers are using more than one method.*

- The key to success is knowing what you want and having a systematic, organized, and informed approach to getting it.

So, while the bad news is: millions of people are looking for jobs right now. The good news (for you) is: most of them have no idea what they are doing.

We want you to be one of the few job hunters who know what you're doing. We have designed our procedure to lift you above the crowd, in order to land you a job.

We believe you will benefit from this book if you fit any of the following categories:

- ☐ **contemplating the possibility of making a change but not sure where to start;**
- ☐ **mildly curious as to whether you will learn anything new about job hunting and career changing;**
- ☐ **desparately looking for a job;**
- ☐ **wanting to help someone who is in transition;**
- ☐ **currently satisfied with your work but would like to be prepared just in case;**
- ☐ **well-read on the subject of job hunting and confused as ever;**
- ☐ **not quite sure why you picked up this book but are experiencing the following:**
  - tiredness, loss of energy
  - detachment from people and events
  - irritability, emotional flareups
  - boredom, cynicism, questioning of life
  - feeling unappreciated
  - limited concentration span
  - psychosomatic complaints
  - depression, disorientation
  - vague discontent with what was once important

In other words, this book is for anyone who sincerely wants to become happily employed. By following the approach outlined in this book, you will soon discover that what may now seem an immense and discouraging task, can become an exciting adventure. Job hunting can provide an opportunity to change and grow, to go places and meet people you never knew existed, to learn about yourself and your capabilities, and to discover you can do things you never dreamed possible (or only considered dreams).

This book is the result of our professional experience and research. We have sifted through the mass of current literature, weeded out redundancies and deadwood and boiled down the rest to the most basic ingredients. We have tested the exercises and techniques with hundreds of clients and proven them highly effective. The result is a simple, concise, and comprehensive overview of the job hunting process, a basic primer, free of excessive verbiage and irrelevant ideas that are rampant in longer, more detailed books.

These pages can be quickly scanned or slowly pondered. You may follow each section closely or refer to passages that fit your needs. Should you desire more indepth treatment of any topic, we offer a reading list at the end of each section. Exercises are included to help you focus and clarify your job objectives and steps are outlined to help you achieve them. We take you from the stage of uncertainty through the focusing until you find the job opening, get hired, and negotiate your salary. We do not leave you without recommending strategies for succeeding on your new job, and pointing out possible income tax deductions. In addition, we have included a special chapter on how to counsel another through the job search crisis.

Our hope is that by reading our workbook, you will easily grasp how to make things happen rather than feel you need to wait for them to happen.

Perhaps you will end up with what you knew you always wanted, or perhaps you will venture into unexpected territory. Whatever you decide, you will travel with the assurance and conviction that accompanies a well thought out decision.

CHAPTER 1
THE JOB-SEARCH PROCESS
A BASIC OVERVIEW

We believe in a step-by-step approach to job hunting. That is not to say the job search is a smooth straight line between start and finish. More often than not the path to employment is a zig zag of false starts, wrong turns, and detours. However, dead-ends need not result in permanent paralysis. When you feel confused or stuck, you can return to a previous step, regroup and proceed from there. Sometimes that even requires going back to the beginning. Yet that doesn't mean starting from scratch again. Enormous progress occurs by eliminating what you know you don't want, or what doesn't work. Therefore, by following such a plan, you may get waylaid, but you should not get lost for long.

Below is an overview of the steps. Look the list over now. Review it when you've finished reading the text. And use it as a guideline when you are actively involved in your job search.

PHASE I:
IDENTIFYING THE JOB YOU WANT

Start. You're confused and unsure. What do you want to do? Where do you start? Your image of yourself is out of focus. What's next? You probably long for the security of a well-defined job and detest the uncertainty of not knowing what's around the next corner.

Focus. With pen and paper in hand, do the exercises in this book and create a detailed checklist of what you want and need in a job.

Develop a job description. Using your checklist create a brief job description.

Informally interview. Describe to others what you want and ask for feedback as to how you come across. Use these mock interviews to refine your presentation.

(If you are changing careers, articulating a new job objective can feel awkward. Repetition helps.) Ask for advice, suggestions and names of other people to contact to build your network.

Now that you have some direction, it is time to find JOB TITLES that fit your description.

PHASE II:
Job Research and Networking

Interview for information. Make appointments with people in positions similar to the one you are interested in and find out exactly what they do on the job. Compare this information to your checklist. Modify your job description as you acquire new information. Note: (at this point) YOU ARE NOT LOOKING FOR A JOB: you're exposing yourself to important leads and information. Do not confuse these purposes or you'll give the impression of vagueness.

Join relevant professional organizations.

Look at trade and company directories and publications in the library.

Read newspapers and periodicals and write to specific companies for their literature.

Explore more than one field. See if your ideal job description matches job options in different industries.

PHASE III:
Getting Hired

Apply for a job. Recontact people you talked with earlier, and let them know you'd like to discuss employment.

Cover all bases: want-ads, agencies, and so on. Don't neglect any approach, but let the time you devote to each activity be proportional to the likely payoff. (remember that 75 percent of all available jobs are never advertised).

Interview for jobs. Get a lot of practice. It takes a string of no's to get the yes! Convince the employer to hire you by selling benefits.

Continue interviewing for information. Continue to gather information about the job market and trends in companies you are targeting.

Consider taking a survival job to fund your job search campaign. It may or may not be a stepping stone in your career direction.

KEEP AT IT WITH DOGGED PERSISTENCE AND BE POSITIVE. You WILL become happily employed.

66 *The self-assessment process helped me to look at my talents and skills and the extent of the mismatch in my previous job.*"

— A former job hunter

CHAPTER 2

PIECING TOGETHER THE PUZZLE

Before you grab your hat and coat and dash out the door to the nearest job interview, you need to take some time to reflect.

There are two approaches to job hunting. One is what we call the Shotgun approach. You're probably familiar with this one, since it's the tack most people take.

It goes something like this: you're desperate to find a job, so you comb the Sunday want ads, send out 765 resumes, knock on every personnel office door within 300 miles, and pray. When someone asks you what kind of work you're looking for, you generally respond. "Anything!" (hoping to keep all fronts covered) or something vague like "Computers -" because Uncle Fred said computers was a growing field—"But I'm open to anything." Time passes. Rejections pile up. Your energy, optimism, and confidence plummet. Things look pretty bleak. Suddenly, luck strikes. The Jolly Janitorial Service needs a bookkeeper. Of course, that isn't quite what you had in mind (not that you really had *anything* in mind except ending the anxiety of being out of work). But at this point, anything looks good. You apply for the job and get it. Time to celebrate, right? Not so fast...

Within a month you begin to grow restless. But you can't just quit. How would it look on your resume? So you stick it out. You drag yourself to work in the morning, watch the seconds tick by on the clock, and dream about becoming a beach bum in Tahiti. You begin resigning yourself to the knowledge that before too long you will have to enter the cold, cruel world of job hunting again. It's either that or stay at Jolly Janitorial forever.

And so it goes. You drag yourself to work day after day until one morning you realize you've *got* to get another job-fast. And you begin the Shotgun cycle again.

A BETTER WAY

Another approach to finding a job is the one we call *FOCUSING.* By focusing, we mean that before you grab your hat and coat and dash out the door to the nearest job interview, you take some time to reflect. You look within yourself and ask yourself some questions.

What could spark your enthusiasm and send you flying out of bed in the morning? What would bring you a feeling of success and satisfaction? What, specifically, are your skills, interests, and needs?

The process is similar to putting together a jigsaw puzzle: if you don't get all the pieces out of the box, you're going to have a tough time completing the picture. In the same way, if you don't uncover parts of yourself that may have been hidden up to now, you'll be unable to form a clear picture of the job that's right for you. These hidden parts include your needs, values, interests, and skills, and they must be clearly identified and evaluated before you can reach your career goal.

THE PROCESS

How do you delve in and come up with all this information about yourself?

One way is simply to ask yourself questions: what am I good at doing? What do I need and value? What do I enjoy? What seems to motivate me? Usually, though, this approach doesn't work very well. It's very difficult for many of us to take stock of ourselves accurately and objectively. Our self-perceptions have been warped over time, and self-doubt distorts our judgment.

Fortunately, however, tools-in the form of questionnaires and inventories-are available to aid in our self-evaluation. These tools range from simple listmaking techniques to sophisticated computer-scored instruments such as the Strong Vocational Interest Inventory. But whatever their level of sophistication, these tools are not crystal balls. They don't operate like slot machines; yielding your ideal job after you've entered your data like so many coins. Nothing-no test, no exercise-will do that for you. There are no magic answers. These tools are mirrors offering you a reflection of who you are, but you will have to be able to see that reflection before you can use it.

This chapter ends with a series of exercises that we have used successfully with hundreds of clients to help them focus and clarify their direction. Each activity is designed to uncover pieces of the puzzle. We urge you to take time out to do these exercises. Only by thinking through your responses, playing with the information, and analyzing the results can you enable ideas to take shape and insights to result.

THE PROCEDURE

When we work with clients, we schedule one-hour sessions twice a week for three weeks to discuss and interpret results of the exercises, which they complete at home. To them-and you-we stress that structure, routine, organization and discipline in performing the exercises are not only helpful; they are vital. We suggest that you set up a schedule for devoting time to self-assessment, and we offer the following suggestions:

1. **Set aside periods of uninterrupted time**, perhaps ninety minutes three times a week, to do the exercises, and read in this book. Stick to this schedule rigorously.

2. **Create a private work space.**

3. **Keep all your written work in a folder or notebook.** (You want to avoid scattering pieces of information all over the place. You probably feel fragmented enough already!) We give our clients folders with side pockets. You will need a similar system for keeping track of records and information.

4. **Keep a small pocket notebook with you** at all times to jot down names, observations, "aha" experiences, and stray ideas.

5. **Discuss your progress with someone supportive.** Great ideas often surface during conversations. Keeping a journal is therapeutic as well as a good source of inspiration.

THE OUTCOME

It is through this soul-searching that you will come up with your checklist of what you want and need in a job. This checklist will serve as the criteria by which you will evaluate your choices and make decisions. The objective will be to find a job that matches this checklist and meshes well with your identity and abilities-the major prerequisite to being happily employed.

Let's take an example. Say that after completing the exercises and analyzing the results, you find that certain factors stand out as important to you:

1. You prefer to be highly visible in the workplace and to have authority over others.

2. You are good at planning projects and instructing and motivating people.

3. You would like your boss to function as a mentor to you, though you would like to retain some decision-making responsibility.
4. You are interested in the financial world and would like to work for a large organization.
5. You would love your own office but want to work in a team atmosphere.

You may object that you already know these things about yourself. We would respond that unless such preferences are written down they usually remain muddled in your head and therefore esentially useless.

By writing down all that occurs to you about your personal job criteria, what may seem like random bits and pieces of self-awareness will form a skeleton of a job description. And by organizing this list into a few sentences, you can clearly and briefly describe what you want in a job. For example, one possible job description based on the above preferences might be: "I am looking for a management position in a financial institution, perhaps a bank, and want to use my skills in planning, training, and human resource development." (Notice that you need not come up with a specific job title at this point, but rather with a brief summary of what you'd like to do.)

By the end of this chapter you should be able to shape your checklist into a clear, crisp job description. As you explore your objectives, you may find that your description fits a number of job titles, or that your goals may change.

We would like to emphasize an important point: Even if you are convinced from the beginning that you want to be a tree surgeon, we urge you to keep an open mind. There is a whole world out there full of exciting and lucrative possibilities that at this moment are unknown to you but may fit your needs.

If you keep your options open, you won't get stuck. Should you reach a dead end, which often happens in a job search, you'll already have identified alternative routes. For example, what happens if you discover that no job openings exist for tree surgeons? Not to worry. Through scrupulous investigation you will have identified another field where you could work with your hands clipping and trimming, creating new shapes, and restoring damaged roots. You become a hair dresser. This may be a silly solution, but it illustrates an important point. Without contingency plans, you risk terminal discouragement on the first time out. With a number of options open-the number being limited by your imagination alone-you'll have plenty of leads to follow when that first disappointment occurs. But without focusing on your ultimate goal, the options you choose might fail to match your personal criteria. Thus, the *FOCUSING* process enables you to hold fast to your objective while simultaneously broadening your scope.

Margie is an example. She came to us bored with her administrative job. After evaluating her work, leisure, and community activities, she realized she needed to be in a more creative position, using her skills in researching, organizing, and analyzing information, and communicating her conclusions both verbally and in written form. She talked with her friends about this conclusion, and they gave her many constructive suggestions regarding where to look and whom to contact. Within five months, she took a job selling advertising for a local magazine, not to meet her original objective but only as a stepping stone to the more creative side of marketing. She made valuable contacts in the sales position, got excellent advice, learned the lingo, and was able to develop a sharper focus in a particular area of interest. Nine months later, she was hired by a small advertising firm as a market researcher. This job matched her criteria perfectly, though she had never even heard of such a position before beginning her exploration.

We offer this example not only to demonstrate the value and process of focusing, but also to suggest that there are many routes to goals-very few are straight, short and clear-cut. The important considerations are that your goals be clear and your efforts be compatible with them.

BENEFITS

Even if you firmly believe that you know precisely what you want, we urge you to engage in the *FOCUSING* process. At the very least, you will find it reconfirming. Plus, by going through these steps now, you will be preparing yourself for the hiring interview, where you will be asked such tricky questions as "Tell me about yourself," "What are your major strengths and weaknesses?" or "Why do you want this job?"

But by far the most dramatic benefit of focusing is that it yields the amazing phenomenon called synchronicity, a term first used by the noted psychiatrist, Carl Jung. Synchronicity occurs once you identify precisely what you want; as if by magnetism, your very intentions seem to pull people and opportunities into your life that support your efforts and propel you toward your goals. We tell our clients to expect coincidences to occur as their goals become more concrete.

Another very important psychological benefit of focusing is its stabilizing effect. No doubt you have noticed that job hunting puts you on an emotiqnl roller coaster. Some days you're surging with optimism and enthusiasm; other days you're overwhelmed with negativity and despair. If you have a checklist and a goal, if you are disciplined and organized, if you feel clear in your own mind about your direction, you are better able to cope with ups-and-down. Focused job hunting feels a little like being caught in quicksand but knowing there's a bottom. It's not comfortable, but at least you know it won't pull you down for good.

Still, periods of doubt, fear, and confusion are a normal, even essential, part of the job-hunting process, and no amount of preparation is going to eliminate the discomforts completely. As with focusing the lens of a camera, you need to go through the blur to achieve a vivid picture. The confusion is part of the blur. The point is not to fight it. If you are unable to ride out the confusion and have become bogged down by negativity, we strongly urge you to seek counseling.

We know from experience that careful self-assessment combined with the practical steps described in the following chapters will produce positive results. Hang in there. When the fog clears and you're doing the work you've chosen, the world will look a whole lot brighter.

Exercise One

PERSONAL PROFILE

This first exercise asks you to begin by dividing your age into thirds. If you are 42, then you have these three age groups: 1-14, 14-28, 28-42 years. Now relax and let your *THOUGHTS DRIFT BACK IN YOUR MEMORY.* We're going to ask you to focus on past achievements that will serve as the foundation for analyzing motivational and career patterns in later exercises. See if you can recall those times in your life when you did something well that you felt very good about and that you thoroughly *ENJOYED* doing. That's what we mean by success. These experiences must be something *YOU* did, not that you observed your brother or friend doing. They may have made big splashes or just tiny, quiet ripples. For example: learning to tie your shoes, reupholstering a chair, finding a job, writing a poem, solving a problem at work, getting married, or planning a party for four or a banquet for 400. The important thing here is that *YOU FELT GOOD ABOUT IT, ENJOYED DOING IT, AND DID IT WELL.*

Try to identify at least three such experiences in each of your three age groups. This may take a few minutes or may require a few days. Don't force anything-just pay attention to whatever surfaces. If nothing comes to mind, that's all right. Give it time. Sleep on it. Perhaps you are being overly critical, or trying too hard to find just the right experience. If you don't try to influence the outcome, if you're willing to take what pops up, then this exercise will be the most effective.

The following chart provides an example of what we'd like you to do once you have recalled nine experiences. We have given you a worksheet divided into three columns. In the **first column,** put your age at the time of the experience, the **second column** is for you to describe in detail exactly what *YOU DID* to accomplish the experience, and on the **third column** write the reason it was a success to you. For example, we may both have made paper doily hats when we were five years old. One person may feel a success because they received compliments. For you, perhaps, the achievement came in doing it all on your own.

PERSONAL PROFILE CHART

Age	Description	Reason
5	My sister gave me the idea to make paper doily hats. Together we collected material scraps, yarn, and bits of paper from around the house, glued them on, and I made about a dozen different designs. We hung them on the wall like in a store and made up a story about a gallery of hats to tell to whomever would listen.	My parents and friends complimented me.
8	My family took a trip to Colorado and took a lot of pictures which I spent a day arranging on construction paper. Then I put the pages into an album and found things about Colorado in an encyclopedia to include too. I took the album to school for sharing day and was complimented by my teacher who put it on display in the hall showcase.	I was recognized for something I created.
10	I gave a speech in class on my favorite food, potato chips, and how they're made. I had persuaded Mom to take me through a potato chip factory to see how they were made. I took pictures at the factory and glued them onto a posterboard to take to class. My teacher gave me an _A_ on the speech because I had spent so much time learning about potato chips and had used good visual aids.	I was recognized for working hard at learning something new.
12	I thought it would be neat for the YMCA camp I went to to have a yearbook for everyone to take home. The Directors liked my idea and agreed to help. I asked the campers to write about what they liked best about camp and interviewed the Staff about the camp's history. It was such a success that the Staff used it to tell parents about sending their kids to camp.	I had created something that was well-accepted by others and was used even after that camp year had ended.
15	My mom and her friend decided to start a catering business. I over-heard them talking about what to call their business, so I looked for names by flipping through magazines, scanning the phone book, and talking with friends. I came up with "PARTY PALS" and my mom and her friend loved it. Mom got a lot of pleasure from telling people that her daughter had created the name.	Mom recognized me for my interest and creativity. I liked being the creator of something that had visibility.

19	A good friend who was a college reporter asked me for help in writing a story on the history of the community college. I started out just gathering information as background material from newspaper clippings and ended up doing the whole thing for him. It appeared on the front page and everyone at the school complimented me. I felt like a celebrity.	I was recognized for my abilities. I liked the celebrity status.
24	I was appointed to head a task force for the Professional Administrators Association I belonged to. Our job was to develop a report on new ways of networking for women that would be presented to the National Convention. I co-authored the report and then presented it at the convention. In the audience was a reporter from a national women's magazine, and she asked to publish the report.	As part of a team, I was able to complete a project and was recognized for my efforts.
26	I was administrative assistant to the President of a safety equipment firm. After looking through trade journals, I noticed that the firm's ads looked dull and uninteresting. One day I found a diplomatic way to mention it to our boss. His response was to come up with some suggested plans for improving our advertising. I met with the Marketing Director and we began talking about our need to make some changes. He asked me to contact a couple of local agencies to come in and meet with him for the purpose of creating a new campaign.	I took a risk and was rewarded for it.

Exercise Two

VALUES INVENTORY

Values are what give meaning and purpose to your life. Below is a list of values. They are essential to your happiness and satisfaction, so it is vital that they be a part of your job choice. Read them and rate them in terms of the degree of satisfaction you would receive from them, using the scale below. Base your reaction on your first gut-level response.

1 - Not very important

2 - Somewhat important

3 - Important, and would like to have in my life.

4 - Very important; critical to include in my choices to success.

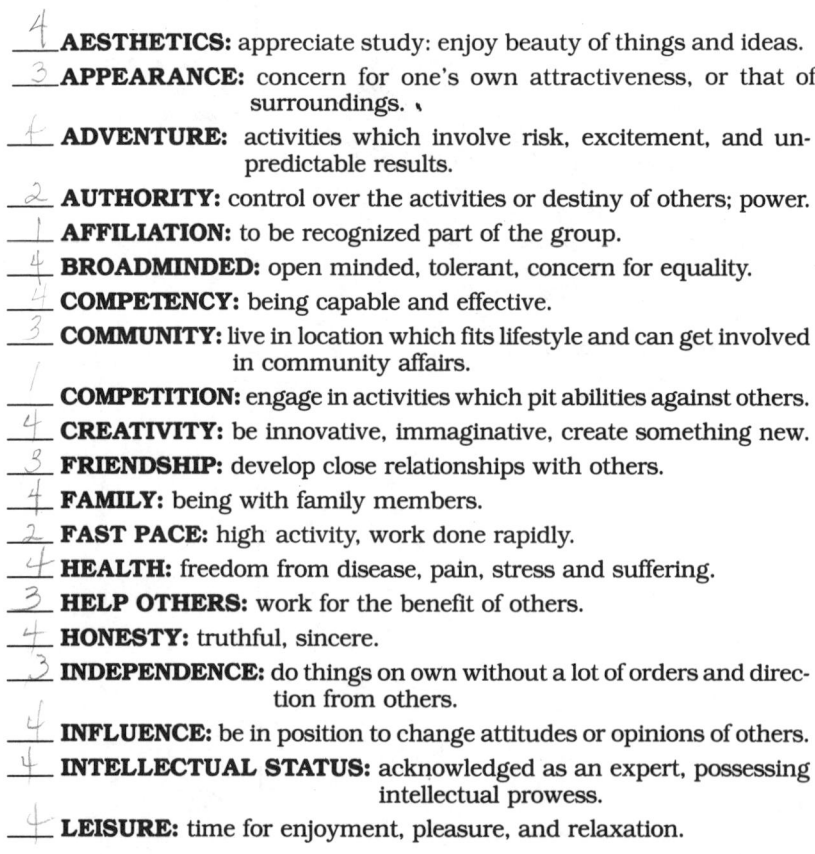

___4___ **AESTHETICS:** appreciate study: enjoy beauty of things and ideas.

___3___ **APPEARANCE:** concern for one's own attractiveness, or that of surroundings. ᵥ

___4___ **ADVENTURE:** activities which involve risk, excitement, and unpredictable results.

___2___ **AUTHORITY:** control over the activities or destiny of others; power.

___1___ **AFFILIATION:** to be recognized part of the group.

___4___ **BROADMINDED:** open minded, tolerant, concern for equality.

___4___ **COMPETENCY:** being capable and effective.

___3___ **COMMUNITY:** live in location which fits lifestyle and can get involved in community affairs.

___1___ **COMPETITION:** engage in activities which pit abilities against others.

___4___ **CREATIVITY:** be innovative, immaginative, create something new.

___3___ **FRIENDSHIP:** develop close relationships with others.

___4___ **FAMILY:** being with family members.

___2___ **FAST PACE:** high activity, work done rapidly.

___4___ **HEALTH:** freedom from disease, pain, stress and suffering.

___3___ **HELP OTHERS:** work for the benefit of others.

___4___ **HONESTY:** truthful, sincere.

___3___ **INDEPENDENCE:** do things on own without a lot of orders and direction from others.

___4___ **INFLUENCE:** be in position to change attitudes or opinions of others.

___4___ **INTELLECTUAL STATUS:** acknowledged as an expert, possessing intellectual prowess.

___4___ **LEISURE:** time for enjoyment, pleasure, and relaxation.

_4_ **LOVE:** devotion, warm attachment, and taking care of loved ones.

_2_ **MATERIAL STATUS:** possessing financial or material possessions.

_3_ **MORAL FULLFILLMENT:** contribute to set of moral standards.

_2_ **ORDER:** neatness, organization, planning.

_2_ **PEACE:** a world free of conflict and war.

_2_ **RELIGIOUS FAITH:** obedience to and activity in behalf of Supreme Being.

_2_ **RECOGNITION:** getting respect, approval, prestige for what you do.

_2_ **SECURITY:** assured of keeping job, free from concern of loss of resources.

_2_ **SELF EXPRESSION:** use of natural talents or abilities which express who you are.

_3_ **WEALTH:** accumulation of money, possessions, property.

_4_ **WISDOM:** mature understanding of life, good sense, and insight.

_____ if you can think of any others, please add.

List below all the values you rated No. 4.

Aesthetics	Family	Leisure
Adventure	Health	Love
Broadminded	Honesty	Wisdom
Competency	Influence	
Creativity	Intellectual Status	

Prioritize the top 5 according to which is the very most important to you, then second most important, etc.

1. Health
2. Family
3. Honesty
4. Adventure
5. Creativity

Exercise Three

PERSONAL PROFILE ANALYSIS

There is a good reason you considered each one of your experiences a success in exercise I. If we disect each story – if we look at what's involved in terms of skills used, relationships with people, maintained, payoffs received, and values listed as important to you – you will see a pattern emerge. In other words, you will begin to see the same skills, values, etc. repeated over and over again. This is called your MOTIVATING PATTERN. This is what turns you on, gets you going, and keeps you stimulated. This same pattern will be the basis of your future choices. For example, you may learn that you are motivated by risk-taking, adventurous situations or encouraging and supporting others.

As you reread each *SUCCESS* experience, answer the questions below and record your answers in the analysis chart provided. Look at the Sample Analysis first as a guideline.

SUCCESS PROFILE ANALYSIS
An Example

1. What skills and abilities did I use to accomplish each achievement? (See page 20.)
2. What kinds of things did I show an interest in?
3. How did I interact with people? What roles did I play?
4. When I look at the reason I considered this an accomplishment, what does this show me that I need and value to give me satisfaction? (See page 14.)

PERSONAL PROFILE ANALYSIS

Achievement	Personality Skills	Transferable Skills/Special Knowledges	Special Interests	Interaction Style/Roles	Needs/Values
#1 Paper hats	creative resourceful artistic	assemble materials develop ideas design	working with objects and things of art	working with another	recognition to be creative
#2 Trip album	creative initiative motivated	organize information arrange materials research	learning about new places	independent	recognition to be creative
#3 Speech on potato chips	resourceful initiative industrious original	present information verbally & visually research explain a process	researching and organizing new information in a creative way	independent with others	to work hard to accomplish a task recognition
#4 Summer camp	innovative ambitious assertive	influence others build a team coordinate a project lead others	researching & organizing information for presentation to others	a leader	authority recognition to be innovative
#5 Catering business	initiative original helpful	identify need investigate create	creative projects	independent a consultant an originator	recognition to be creative to be helpful to others to be visible
#6 Write a story	resourceful reliable motivated	research organize interview journalistic writing	historical research	independent a colleague	to be visible peer recognition
#7 Head of task force	goal-directedness self-confident verbal	develop report delegate tasks analyze information write public speaking	working for the benefit of women's issues	a leader a team player an idea developer	achievement to be part of a team effort - affiliation influence authority
#8 Administrative assistant at the safety firm	assertive resourceful creative motivated	identify a need consult create recommend	advertising/ marketing	independent a consultant	responsibility - authority acceptance of ideas competency

SKILL CATEGORIES

It is often very difficult to identify our skills because of our tendency to lump all our abilities, dismiss certain skills thinking "Anyone can do that," or completely ignore some important assets.

Perhaps it would help to look at skills as falling into three categories:

1. **PERSONALITY SKILLS** (Otherwise known as SELF MANAGEMENT SKILLS) also known as "personality traits" since we rarely think of them as skills. These skills have to do with how you deal with people and time. To get in touch with your Personal Skills, ask yourself this question: WHAT PERSONAL CHARACTERISTICS DO I HAVE? Look at these examples:*

punctual	imaginative	patient
dependable	industrious	persistent
conscientious	leadership ability	sincere
congenial	loyal	athletic
efficient	openminded	sincere

WORK/ACTIVITY SKILLS Specific skills related to performing a job/activity in a particular field or work situation or subject matter which are picked up as one goes through life. They involve learning and memory. To get in touch with your Work Skills, ask yourself this question: WHAT SORT OF SPECIFIC KNOWLEDGE DO I HAVE? Here are a few examples to get you started:

programming computers	pumping gas
knowing anatomy	preparing a lesson plan
understand how to crochet	bowling
typing legal document	repairing a car
tailoring suits	organizing closets

TRANSFERABLE SKILLS Skills needed to deal with data (information), people, and/or things. They are not specific to any particular task but can be generalized to a wide variety of settings. To get in touch with your transferable skills, ask yourself: WHAT KINDS OF ACTIONS WOULD I MOST WANT TO USE IN VARIOUS ACTIVITIES? For example:*

DATA	PEOPLE	THINGS
synthesizing	negotiating	precision work
coordinating	supervising	setting up
analyzing	consulting	kanipulating
computing	treating	operating
compiling	instructing	driving control
copying	persuading	handling
comparing	diverting	handling
	exchange information	tending
	taking instruction	
	helping	

*See following pages for a more complete list.

Examples of
PERSONALITY SKILLS

Following is a more complete list of personality skills which can be useful in assessing one's strengths.

academic	dignified	liable	resourceful
active	discreet	logical	responsible
accurate	dominant	loving	retiring
adaptable	eager	loyal	robust
adventurous	easygoing	mature	self-confident
affectionate	efficient	methodical	self-controlled
aggressive	emotional	meticulous	sensible
alert	energetic	mild	sensitive
ambitious	fair-minded	moderate	serious
analytical	farsighted	modest	sharp-witted
artistic	firm	motivated	sincere
assertive	flexible	natural	sociable
attractive	forceful	obliging	spontaneous
bold	witty	open-minded	spunky
broadminded	forgiving	opportunistic	stable
businesslike	formal	optimistic	steady
calm	frank	organized	strong
capable	friendly	original	strong-minded
careful	generous	outgoing	supportive
cautious	gentle	painstaking	sympathetic
charming	goal-directed	patient	tactful
cheerful	good-natured	peaceable	teachable
clear-thinking	healthy	persevere	tenacious
clever	helpful	pleasant	thorough
competent	honest	poised	thoughtful
competitive	humorous	polite	tolerant
compulsive	idealistic	practical	tough
confident	imaginative	precise	trusting
conscientious	independent	progressive	trustworthy
conservative	individualistic	prudent	unaffected
considerate	industrious	punctuality	unassuming
consistent	informal	purposeful	understanding
cool	initiative	quick	unexcitable
cooperative	intellectual	quiet	uninhibited
courageous	intelligent	rational	zany
curious	introspective	realistic	verbal
daring	inventive	reflective	versatile
deliberate	kind	relaxed	warm
dependable	leisurely	reliable	wholesome
determined	light-hearted	reserved	wise

WORDS TO DESCRIBE TRANSFERABLE SKILLS

Here are some action words to use to describe skills that can be transferred from one job to another.

accomplish	create	index	promote
achieve	decide	influence	propose
act	define	inform	provide
adapt	delegate	initiate	publicize
adjust	demonstrate	innovate	publish
administer	design	inspect	recommend
advertise	detail	install	reconcile
advise	determine	institute	record
affect	develop	instruct	recruit
analyze	devise	integrate	rectify
anticipate	direct	interpret	re-design
apply	distribute	interview	relate
approach	draft	investigate	renew
approve	edit	invent	report
arrange	educate	lead	represent
assemble	encourage	maintain	research
assess	enlarge	manage	resolve
assign	enlist	manipulate	review
assist	establish	market	revise
attain	estimate	mediate	scan
author	evaluate	merchandise	schedule
budget	examine	moderate	screen
build	exchange	modify	select
calculate	execute	monitor	serve
catalog	expand	motivate	speak
chair	expedite	negotiate	staff
clarify	facilitate	obtain	standardize
collaborate	familiarize	operate	stimulate
communicate	forecast	organize	summarize
compare	formulate	originate	supervise
conceive	fund raise	participate	survey
conceptualize	generate	perceive	synthesize
conciliate	govern	perform	systematize
conduct	guide	persuage	teach
consult	handle	plan	team build
contract	hire	present	train
control	identify	preside	transmit
cooperate	implement	problem solve	update
coordinate	improve	process	utilize
counsel	increase	produce	write

Exercise Four

CLUSTERING YOUR SKILLS

Now we can start grouping or arranging what we learned from the Personal Profile Exercises. Take a look at the skills you wrote down then group your skills into categories listed below. You will find that some skills will fit into several clusters (i.e. "can communicate with management" could fall into managerial as well as verbal communication). If you cannot decide which cluster to include the skill under, list it under both to eliminate unnecessary confusion at this point. Don't worry about being perfect-just do it.

Seen below are the skill families and some examples for each:

LEADERSHIP
lead, motivate, enthusiastic

TEACHING/TRAINING
design, present

WRITING/EDITING
write, edit brochures and newsletters, organize

PHYSICAL/OUTDOOR
coach volleyball team

ARTISTIC/PERFORMING
graphic design, creative writing

SCIENTIFIC/MATHEMATICAL
calculate, knowledge of statistics

ADVOCACY
identify areas for change, develop and implement an action plan

MANAGERIAL
organize, implement, interview, understand management by objectives, supervisory abilities

VERBAL COMMUNICATION
able to communicate concepts, public speaking

OFFICE/MANUAL SKILLS
operate office equipment

RESEARCH/ANALYSIS
resourceful, develop ideas

SOCIAL/INTERPERSONAL
counsel individuals, sensitive, nurturing

SALES/PROMOTIONAL/ FUNDRAISING
market/sell products, conduct a market survey

YOUR OWN CLUSTER

_____

_____

_____

Sample Form

CLUSTERING YOUR SKILLS

LEADERSHIP

_____
_____
_____

MANAGERIAL

_____
_____
_____

TEACHING/TRAINING

VERBAL COMMUNICATION

WRITING/EDITING

OFFICE/MANUAL SKILLS

PHYSICAL/OUTDOOR

RESEARCH/ANALYSIS

ARTISTIC/PERFORMING

SOCIAL/INTERPERSONAL

SCIENTIFIC/ MATHEMATICAL

SALES/PROMOTIONAL/ FUNDRAISING

ADVOCACY

YOUR OWN CLUSTER

Exercise Five

FEEDBACK SHEET

A critical, but much neglected tool in making career decisions is feedback from other people. We want to show you an easy and fun way for you to get it. Make four copies of this sheet. Give a copy to friends, business associates, family, etc. (any one with whom you would feel comfortable doing this), and ask them to fill this out. Don't look over their shoulder while they do this, but you may want to discuss their comments afterwards. You should learn a lot and be delighted with the results! Choose a variety of individuals who know you in different ways.

1. What do you see as my major personality strengths?

2. What do you see as my most marketable skills?

3. What kind of environment do you see me working in?

4. What do you see I need in a job?

5. What do you see in me that I probably don't see in myself?

Exercise Six

MAGIC WAND (I)

Now it's time to really have fun. In order for this exercise to be effective, you must be totally fanciful and even a little crazy!

We're going to give you a magic wand. If with a flick of the stick, you could have anything you wanted in a job, what would you include? Make a "grocery" list below of everything you would ideally love to have on your job. Don't try to be realistic –that comes later. Right now just put down what you would like to have in your ideal job. When you can't think of what else you'd like, list everything you know you would *DEFINITELY NOT* want.

> **EXAMPLE:** *Money–$25,000 to $35,000*
> *Travel–about 8 to 10 trips a year, even*
> *overseas.*
> *Opportunity to speak in public–preferably*
> *large groups of 50 to 100, etc.*

Remember, you don't have to show this to anyone; nor will you necessarily ever bring all of this into reality. The point is to have fun, and write whatever comes to mind. Go on - you can do it!

MAGIC WAND (II)

You still have this magic wand. Now, from your "grocery list" of ideas, write a little story, a description of your ideal fantasy job. DO NOT INCLUDE A JOB TITLE!

You may think this exercise is silly or senseless. However, we believe tucked in this ideal job is a large kernel of truth. It represents your dream, your long range goal, that to which you truly aspire. Respect it. This reflects very real needs in you. We even suspect the basic ingredients will be similar to those found in your Personal Profile Analysis.

We want you to use this "dream criteria" to evaluate job choices. As you research employment opportunities, ask yourself, "How much of my criteria will be met in this job?" Of course, you will have to be realistic, prioritize your needs, and sacrifice some for the sake of others. But, to be happily employed, you must never lose sight of that dream.

EXAMPLE: *I would go to a rather large company, and I would be, if not the president, then next in command. It would be a very important position with a lot of clout. I would spend the morning alone, doing research, completing assignments, doing a lot of writing. In fact, I could see working for a publishing company. I would lunch with friends or colleagues. We would have weekly staff meetings where we would brainstorm ideas. My work would get published, I would be interviewed on T.V., and I would be called on to speak at gatherings all over the country.*

Exercise Seven

SUMMARY SHEET

By now you have a lot of valuable information written down. You have looked at your past achievements and picked out skills, interests, values, and interactive styles. In addition, you have prioritized your needs/values, clustered your skills, obtained feedback, and even taken a flight into fantasy with your magic wand.

Now, the question is, "What do you want to do with this mass of information?" Do not worry. We want to help you pull everything together. The chart below is a good organizational tool.

Based on what you discovered in the previous exercises and everything you know to be true about yourself, fill in each section with the appropriate information.

NEEDS/VALUES	SHORT/LONG TERM GOALS	FINANCIAL NEEDS
1. Authority 2. To be creative 3. 4. 5.	1. Invest in real estate 2. 3. 4. 5.	1. 20,000 plus 2. Pension Plan 3. 4. 5.

WORKING ENVIRONMENT	INTERESTS	WHERE GEOGRAPHICAL	PERSONALITY SKILLS
1. Happy, supportive people 2. My own office 3. 4. 5.	1. Design programs 2. Work with my hands 3. 4. 5.	1. East coast 2. Close to home 3. 4. 5.	1. Motivated 2. Dependable 3. 4. 5.

SPECIAL KNOWLEDGES	TRANSFERABLE SKILLS	ROLES
1. Organizational development 2. Computer programming 3. 4. 5.	1. Manage 2. Teach 3. 4. 5.	1. Leader 2. Teacher 3. 4. 5.

Exercise Eight

JOB DESCRIPTION (I)

This sheet helps you to condense, think through, and begin forming a job description.

Fill in the following statements. Again, don't try to be perfect. Just write down whatever your heart feels is right.

I see that my major SKILL areas are _____

_____

_____

My skills tend to be more related to ___ PEOPLE ___ DATA ___ THINGS.

The skills that are important to me, and that I most ENJOY using are _____

_____

Perhaps some strengths and skills I may want TO DEVELOP include _____

_____

I see that my INTERESTS are _____

_____

_____

Perhaps I would like TO DEVELOP the following interests

_____

_____

I found I like TO INTERACT with people by _____

_____

_____

I function best in ROLES where I can act as a _____

_____

_____

I learned that to be happy and fulfilled, I NEED _____

_____

_____

I was SURPRISED that _____

_____

_____

I learned that I am MOTIVATED by _____

_____

_____

Some IDEAS I have from doing this are _____

_____

_____

Exercise Nine

JOB DESCRIPTION (II)

A job description is a brief 30 second response to the question, "What kind of job do you want?"

This worksheet will help you form an answer to that very question. Remember, this is not a specific position, but an objective which may match several job titles.

Begin by completing the following statements. Just write. Don't worry about what it sounds like now. You will polish it up later.

The kind of position you want: (eg. director, sales person, coordinator, member of a team, manager, consultant)

Using these skills: (those strongest skills and abilities you have already identified)

In this kind of organization: (eg. small, large, nonprofit)

Solving these kinds of problems: (eg. financial problems, overcoming lack of public awareness)

To achieve these results: (eg. increase profit, educate public, create harmonious environment)

Exercise Ten

JOB DESCRIPTION (III)

Now, write a paragraph combining the statements you wrote in the previous Exercise (Job Description II). It will be long and perhaps a bit awkward at first. However, you will need to rewrite it several times. It may be helpful to say it aloud as you write.

Sample Job Objective 1:

I am seeking a managerial position in a small to medium size company where I can use my skills in administration, training and marketing to motivate the sales staff and create new marketing strategies for implementation.

Sample Job Objective No. 2:

I am seeking a staff position in a community agency which could integrate my skills in project coordination, volunteer supervision and promotion to improve the community image of the agency and promote its causes.

My Job Objective

_____

_____

_____

_____

_____

There, you did it! A clear, focused description of the kind of job you want. It may change (and probably will), but at least you have a direction and a starting point.

OTHER TOOLS OF SELF-ASSESSMENT

To help you draw more pieces of the puzzle out of the box, we strongly recommend these additional tools of self-assessment.

Testing and Analysis

Personality, aptitude and vocational tests vary in format and degree of sophistication. Some are multiple choice, some are sentence completers, and others can be questionnaires. The majority of these tests have been designed to assess personality, interests and aptitudes for a particular career.

We want you to keep in mind that tests are merely indicators of vocational preference and should serve only to reconfirm the information gained through self-assessment. In other words, testing is another method for piecing together the puzzle; alone it is not as valid.

The standardized tests described here should always be administered and interpreted by a trained psychologist or career counselor who is knowledgeable about the instruments and their applications as well as limitations. Testing is usually available from a variety of career consultants, psychologists and college placement and community centers.

The **Strong-Campbell Interest Inventory** (SCII) is the most commonly used vocational test. It gathers information about the test-taker's interest patterns and suggests possible occupations based on the specific work activities involved, the characteristics of the working environments and the personality traits affecting the work. The test-taker's interests are compared with satisfied workers in about ninety different occupations and the degree of similarity or dissimilarity is scaled.

The **Edwards Personal Preference Test** is a forced choice instrument which is used to clarify needs and values of the test-taker. The test ranks fifteen different values on a percentile and individual basis which can be directly related to personal work preference.

The **Myers-Briggs Type Indicator** is a work preference indicator. This multiple choice test reports a work preference profile which can be related to types of compatible people, environment and work activities.

The **General Aptitude Test Battery** measures a range of aptitudes such as verbal and numerical abilities, manual dexterity, mechanical abilities, spatial relationships, etc. This is a timed test and is usually administered in a small group. The results are especially useful for vocational re-entry and re-training.

The Tummy Grabber

We found this to be a helpful and fun tool for "visual" people. During the next 4 to 6 weeks, collect articles, pictures, advertisements, etc. that "grab" you. Anything you see, hear, or read that peaks your interest, that causes you to react positively, tear out (or jot down a few words about it on a piece of paper) and toss it in the envelope. Do not try to figure out why you chose this particular item, just collect the material.

At the end of 4 to 6 weeks, lay the material out on the floor and see what you have. You may be in for some surprises. Do you pick up any patterns? Do you see anything new popping out? Or perhaps this serves to reinforce what you already suspected was your interest.

Begin today. Do not analyze--just tear out the material, put it in the envelope, and forget it.

SUGGESTED READING
Chapter 2

Bolles, Richard N., **What Color Is Your Parachute?** *Berkeley, California. Ten Speed Press.* An excellent manual for anyone engaged in job hunting or career change. Exercises in self-assessment as well as tips for job hunting and extensive lists of resources. Revised annually.

Figler, Howard E., **The Complete Job Search Handbook.** *New York. Holt, Rinehart, and Winston, 1979.* Tools and exercises to inventory your marketable skills and prepare you for a successful job hunt.

Haldane, Bernard, **How to Make a Habit of Success.** *Washington, D.C., Acropolis Books, Ltd., 1960. Considered to be one of the best books on career/life planning.*

Miller, Arthur F. and Mattson, Ralph T., **The Truth About You: Discover What You Should Be Doing With Your Life.** *Tappan, New Jersey, Fleming H. Revell Co., 1977.* Based on thousands of case studies, this book will help you identify your basic motivational patterns and teach you how to apply these to your total life –career and family.

Sher, Barbara, **Wishcraft: How To Get What You Really Want.** *New York, Viking Press, 1979.* Her exercises and personal anecdotes help you to examine your career dreams and how to make them reality.

Especially For Women

Catalyst Publications, 14 East 60th St., New York, New York 10022. Publishes a multitude of self-guidance, career-related materials for women.

Scholz, Nelle Tumlin; Prince, Judith Sosebee; and Miller, Gordon Porter, **How to Decide, A Workbook For Women.** *New York, Avon Books, 1978.* Helps you assess where you are and who you are. Provides information and strategies for making decisions and following through with an action plan.

66 *It's who you know . . . and, once*
they know what you know, you're
well on your way to finding those
hard-to-find opportunities."

— A former job hunter

CHAPTER 3

INFORMAL INTERVIEWING

When you know what you want and can communicate your goals to others, you are well on the way to becoming happily employed.

What if you have come up with a marvelous sounding job but have no idea what to call it? What if no such position exists as far as you know? Don't be too concerned. The purpose of this chapter is to help you find job titles that fit your criteria. Also, by matching your goals to jobs in the outside world, you will clarify your objectives. Finally, in the process of clarification, you will be building and expanding your network. How? When you are able to clearly tell other people what you are looking for, they will be able to link you up with just those people who can further your plans. This is called networking. And since 75 percent of all jobs are found via the grapevine, networking is what job hunting is all about.

However, for some people, the thought of approaching people for help has about as much appeal as walking barefoot through a field of broken glass! If you feel this way, give us a chance to convince you that networking for a job can be painless and even fun. "What! Fun?" you retort, rolling your eyes. "Are you out of your mind?" Wait and see: you'll find that if you are focused and proceed step by step, you can enjoy yourself *while* you're working toward your goal. Stay with us here; let us show you.

To be effective, networking has to be done right. For convenience, we have broken down the networking process into two phases, one covered here and the second in the next chapter. The first phase is what we call *INFORMAL INTERVIEWING*. Informal interviewing means talking to people you know (friends and family) in order to obtain information, advice, assistance, feedback, and support. Done correctly, this kind of interaction is a low-stress but highly effective entry into the job market.

THE OUTCOME

Informal interviewing can yield a wealth of benefits. Like dress rehearsals before opening night, this process allows you to practice your act without the pressures of having to perform perfectly. It's easier, and wiser, to test the waters with a friend than with a stranger who will decide your career fate. What exactly are the benefits? Above all, you'll have the chance to polish your presentation and further clarify your objective.

1. Each time you express your goal **you become more focused,** articulate, and confident. In the beginning, and especially if you are making a career switch, you may feel awkward in describing your new ideas. It is often very difficult to see yourself in a different role. And, as we mentioned, you will probably find yourself modifying, and even changing, your original objective as you learn more about yourself and the world.

2. **You'll receive valuable feedback** on how you come across. How are other people reacting to you? Are they confused by what you're saying? Are you demeaning or disqualifying yourself unconsciously? Find out what areas in your presentation you need to refine in order to make a good impression when it counts. Also, when the feedback is positive and encouraging, it will fill you with confidence and energy.

3. **You'll get new ideas about how to realize your goals.** Upon hearing your job description, your listeners might know people who do the work you describe. In this way, you'll learn job titles as well as names of organizations, fields, and industries in which your skills and aspirations are valued.

4. **You'll learn the names of people working in jobs similar to the one you describe.** This information will lead to the second phase of networking, informational interviewing. Only by talking to people who are actually doing what you are considering, can you (a) judge whether a certain field is right for you, (b) make important and valuable contacts, and (c) uncover those unadvertised jobs. Consider how much easier it will be to call a stranger when you can say "Your cousin told me I should talk to you" than when you have no personal link at all.

PREPARING

To get the most from your informal interviews, you'll need to take the following steps:

1. **Have business cards printed** to give to those you interview. This is a must. Any print shop will make up simple cards showing your name, address, and telephone number. People save cards. Scrawling your personal information on

the bottom of your grocery list or a check stub and tearing it off to give someone is not only messy, but it looks tacky. A major goal in these interviews, besides collecting information, is to project a professional image that reflects the seriousness of your intentions. Equally important in this regard is dressing appropriately. No matter whom you are meeting, wear businesslike clothes to set the tone.

2. **Have letterhead stationery printed** with your name and address, and telephone number in a neat businesslike typeface.

3. **DON'T ASSUME ANYTHING.** If you say to yourself that "there are no jobs out there like this." or "no one is going to hire me," you'll be creating self-fulfilling prophecies. False assumptions account for many missed opportunities. Wait until you have checked out your idea thoroughly before coming to conclusions.

4. **Remain focused.** At the risk of belaboring the point, we stress again the fundamental importance of identifying and concentrating on your goals. To seek out help that is relevant, and to avoid flying off in a thousand different directions, you must first know what you want. Your ability to articulate a job description will help others to see what you are after.

5. **You may not want your current employer to know that you are looking for a new job.** If this is the case, you can still discreetly engage in informal interviewing as long as you remember to ask your contact to keep all discussions confidential.

NOTE: Even if you have zeroed in on a job title, we encourage you to do the informal interviewing and leave yourself open to other possibilities. By clinging to the familiar, or by deciding too quickly, you may overlook exciting, fulfilling, and profitable alternatives to your original choice. Just because you have only worked as a teacher doesn't mean the word "teacher" is indelibly stamped on your forehead. Changing your professional identity is partly a matter of repackaging yourself. Informal interviewing is part of the packaging process.

THE PROCEDURE

Very briefly, informal interviewing includes the following five steps:

1. **Make a list of people you know personally.**
2. **Meet with them to discuss your career plans.**
3. **Ask them for ideas, resources, leads and feedback.**
4. **Keep written records of all suggestions and your action taken.**
5. **Follow up all informal interviews with thank-you letters.**

1. MAKE A LIST OF EVERYONE YOU KNOW

When we ask our clients to write down the names of everyone they know, they look at us in dismay. "Everyone?" they exclaim, discouraged by the enormity of the task. However, the job isn't as tedious or as unwieldly as it might seem at first. When you start writing, it goes quickly.

Include only those people with whom you feel comfortable talking. Just start listing friends, family, neighbors, school chums, army buddies, and people you do business with, such as your accountant, stockbroker, hairdresser, dentist, and so on. Make sure you include those in sales or service jobs whom you know casually; they are sure to have a lot of contact and community awareness. For example, one woman, while her dress was being altered, described to the tailor her dreams of being in the travel business. As it happened, the tailor's next client owned a travel agency and came in bemoaning the loss of an employee during a particularly busy period. The tailor gave the travel agent his previous client's card, and a successful match was made.

Don't rule out acquaintances in other cities, either. Another job hunter who had recently moved to our area wrote to friends back home and received a pile of return letters full of wonderful feedback, encouragement, ideas, and even names to contact in his current location. As an added advantage, the writing gave him an opportunity to define his objective.

Keep prodding yourself to make this list comprehensive. Really stretch your memory. You may not need to contact more than a few of the individuals, but invariably, the last names on the list will prove to be the most fruitful contacts.

With such a list in hand, you never have to feel stuck, since, at any point in your job search, you will always have this list to come back to.

Consider Jon, who became very discouraged when, after three

months of exploring computer sales, he realized that it was just not a field for him. Referring to his informal interview sheet, he recontacted friends to get some new ideas. This resource made it easier for him to get through a distressing period.

2. ARRANGE APPOINTMENTS

Look at your list and study your calendar. How many people could you see this week? Two? Three? Four? Check those individuals you would be willing to call and with whom you would feel comfortable discussing your job hunt. Did you check with your father? Uncle Fred? Will you tell your dentist with whom you have an appointment tomorrow (that counts)? Initally, choose only those whose company you enjoy regardless of their connection. Studies reveal that on the average, every adult knows 250 people. Thus, every contact you make potentially exposes you to a myriad of possibilities–that person's own 250 contacts. Therefore, disqualify no one as a possible source of help. Exciting leads come from unexpected sources.

It is very important to set face-to-face meetings, as opposed to telephone conversations. Personal contact is infinitely more productive. To set this date, you might say something like, "Aunt Sara, I am thinking about changing jobs. I'm not quite sure what I want to do, but I have been playing around with some ideas, and I'd love to hear your advice. Could we meet whenever it's convenient for you ... perhaps for lunch this week?

3. MEET WITH YOUR CONTACTS

At the meetings, briefly describe your objective. Pay attention to your contacts' reactions. Do their eyes glaze over halfway through your first sentence? Learn where and when you need to polish your presentation. Every time you repeat your ideas you will find yourself becoming more confident and clear. Remember, these first attempts to describe new goals may feel clumsy, even embarrassing. That's why it's better to practice and experiment with a trusted friend than with a potential employer. Also, always ask, "Who do you know who may be doing what I am describing?" or "Can you think of anyone else you think I should talk to?"

4. KEEP RECORDS

Write down the names, addresses, telephone numbers of any referrals; note the person who gave them to you. After many discussions, you may forget who referred you to whom. Also, jot down any ideas and suggestions.

5. SEND THANK-YOU NOTES

That's right – even send a note to Uncle Fred. A little note of appreciation is all that's necessary. It's a nice gesture in itself and is also an important habit to start cultivating in the interest of maintaining business contacts.

BENEFITS

Once these informal conversations have helped to build your resource network, your job search is well under way. And since you've been dealing strictly with people you know, the process shouldn't have hurt a bit. Far from writhing with the old job-search anxiety, you may even have enjoyed yourself by enriching old relationships and doing a bit of socializing. Your contacts, on their part, have undoubtedly felt flattered at being sought out for advice.

It's inevitable that you will experience some seemingly wasted encounters. But for every two, three or four pointless discussions, you'll have had one yielding enormous reward. It is the law of averages: the more people you talk to, the greater will be your chances of success.

ONE FINAL NOTE

It's a good idea to read about different occupations that interest you, and about the job-hunting process in general, as you pursue your networking efforts. Visit a public library or college career center. The federal government has some excellent material available through the Superintendent of Documents or the federal bookstores. Ask them to send you some literature. In particular, the *Dictionary of Occupational Titles,* and the *Occupational Outlook Handbook* are excellent general references for specific fields and are available in most libraries. (Other good sources of career information are included in the chapter bibliography.)

Keep your ears and eyes open. Be a sponge, soaking up new information and ideas so you will be ready for the next step– *INFORMATIONAL INTERVIEWING.*

INITIAL CONTACT LIST

Use this form to brainstorm and list people who could be a part of your initial network.

WHO: Anyone you know: family, friends, neighbors, former business associates, teachers, etc.;
Anyone who provides a service and deals with people constantly (e.g. bankers, lawyers, politicians, clergy, civic leaders);
Anyone who earns a living by making contacts (e.g. sales people, stockbrokers, real estate people, insurance, etc.).

WHY: To get names of people doing what you are interested in doing;
To get their initial reaction to your career goals and your presentation;
To get practice in interviewing and asking questions;
To get ideas.

HOW: Set an appointment, tell them what you are thinking about, and ask for names, ideas, and suggestions.

names	names

MARKETING YOURSELF IN PRINT

Effective colors for stationery and business cards are grey and ivory. Consult with a typesetter or printer as to style of type, paper quality and color. This is an important investment for job hunters, as it reflects your professionalism and personal style.

Example of
Business Stationery

Jeanie Smithson
11200 Alhambra Drive
Kansas City, Missouri 64114

Jeanie Smithson
11200 Alahambra Drive
Kansas City, Missouri 64114

Example of
Business Cards

Jeanie Smithson
11200 Alahambra Drive
Kansas City, Missouri 64114
(816) 361-7198

ORGANIZING THE RESEARCH

Record-Keeping: After each interview for information and feedback, complete a record like the one here.

Contact File: 5x7 cards or pages in a notebook.

Name _____ **Position** _____

Company/Organization _____

Address _____ **Referral source** _____

Phone _____

Type of Communication (letter, phone, interview) _____

      Date _____ **Content** _____

      _____

Referrals, Suggestions _____

_____

Follow-up Strategy _____

_____

Resume sent _____ **Date** _____

SUGGESTED READING
Chapter 3

Germann, Richard, and Arnold, Peter, *Job and Career Building.* *Berkeley, California, Ten Speed Press, 1982.* Excellent for explaining the informal interviewing process as well as the whole job search. You may need to order this one, but it is worth it.

Kisel, Marie, *Design for Change: A Guide to New Careers.* *New York, Franklin-Watts, Inc. 1980.* Explores career options in major occupational fields and specifically discusses career changes from areas such as education and government.

Welch, Mary Scott, *Networking.* *New York, Warner Books, 1981.* A basic primer on the how-to of networking.

Government Publications

Occupational Outlook Handbook. Gives the outlook for major occupations as well as other important information about qualifications, expected earnings and related job titles.

Dictionary of Occupational Titles. Lists thousands of occupations according to the U.S. Govenment classifications system. Available from the Superintendent of U.S. Government Printing Office, Washington, D.C. 20402; the public library or any federal book store.

Other Resources

Professional and Trade Associations. Most will send you free literature about their profession or trade. Many publish journals.

In the library, locate:
Directory of National Trade and Professional Associations in the U.S.
Standard Periodical Directory.

U.S. Department of Commerce. Can guide you to information about specific industries and companies. Write to: Bureau of Industrial Economics, U.S. Department of Commerce, Room 4845, Washington, D.C. 20230.

Other Publications

Valuable resources for learning about industries, employment outlook and specific companies.

New York Times   For Women: **Savy**
Wall Street Journal   **Working Woman**
Barron's Weekly

CHAPTER 4

INFORMATIONAL INTERVIEWING

Becoming happily employed involves more than just knocking on doors in search of an empty slot.

Finding the right job is a process of discovery and experimentation. It involves continually learning about yourself and the world and clarifying, modifying, and refining your goals until you have hit upon a compatible and available niche. As we mentioned earlier, the key to expediting this process is networking. Informal Interviewing, discussed in the preceding chapter, was the first phase. Here we take you one step further into researching your identified job titles.

The best way to learn more about particular jobs and gain a competitive edge is through INFORMATIONAL INTERVIEWING, the second phase of networking. Simply put, informational interviewing involves meeting with people in jobs you would like to have as well as talking to others in your field of interest who may be potential employers. You ask them questions about what they do and about the trends in their trade or profession. It's just what the term implies-you interview them for information.

THE OUTCOME

What's the purpose of INFORMATIONAL INTERVIEWING? These interviews give you an ingenious, stress-free opportunity to maneuver yourself into the right place. Under the guise of research, you are able to observe an organization in operation. At the same time, you can present yourself to a potential employer without either of you feeling any pressure to make a decision. Richard Bolles, in his book, *What Color Is Your Parachute?*, compares this approach to window shopping, and says, "Any way you can let an executive window shop you, without putting them on the spot, will create a very favorable situation for you." Simultaneously, you will be able to window shop for the most appropriate working situation.

By interviewing people in the field, you can reap the following benefits:

1. **You can find out before you make a commitment** if the job fits your criteria.

2. **You can get your foot in the door.** When you request an informational interview, and not a job interview, a person will be more apt to meet with you. (Even more appealing to them will be your interest in what is likely to be their favorite subjects: themselves and their work.)

3. **You will be adding to your network** by making contact with people in the field. Remember, the name of the game is who you know.

4. **Your efforts and thoroughness in job hunting are likely to impress your contact** who may one day be in a position to hire you.

5. **You'll be by-passing the personnel office,** which as a rule has no hiring power, but exists only to screen applicants. Why risk disqualification before you can get to the person with authority?

6. **You'll have an opportunity to learn the problems and needs of the industry** as well as its jargon, which will work to your advantage in the hiring interview.

7. **You'll have a chance to find out about unadvertised openings** or gather ideas where new positions might be created.

8. **If you want to work as an independent contractor,** you'll have a chance to talk to those who have followed a similar career path or who have worked with free lancers. They can serve to motivate you and help you create a successful strategy.

There's only one catch. If you're going to do this right (and why do it if you're not), you must follow the first commandment of Informational Interviewing: *THOU SHALT NOT ASK FOR A JOB: THOU SHALT ONLY SOLICIT INFORMATION.*

Violating this rule will damage your efforts as well as your credibility. In short, break the rule, and you could blow everything.

At this point in your job search, you are not looking to be hired but to learn more about a certain area in order to make a decision. True, a job might well result from your contact with people who have the power to hire, but, at this point, you are not yet ready to make a decision and seek a hiring interview.

It's important that this distinction be clear in your mind. If you are confused about the purpose of your visit your interviewee will be confused, too, and you will wind up making a very poor impression.

And if, deep in your heart, you don't give a hoot for the research and are really after a job offer, then for heaven sakes say so. Pretending otherwise could seem like deceit.

We hope, however, that you will appreciate the value of the informational interview not only as a means of learning and of making contacts, but also, should you decide you would like to return, you'll have an advantage over an applicant who is entering the premises for the first time. If, however, in the course of conversation, you are offered a job – which often occurs – it is advisable not to accept on the spot. For your own benefit, take some time to think it over.

PROCEDURE

Informational interviewing consists of four steps:

1. **Either by phone or letter, arrange a face-to-face meeting.**
2. **Meet with the interviewee to ask questions and obtain information.**
3. **Follow up with a thank-you note.**
4. **Maintain careful records of all communication and follow-up.** (Use the contact file cards suggested in Chapter 3.)

More specifically, we suggest the following:

1. CONTACT PEOPLE WHO ARE IN JOBS OR ORGANIZATIONS THAT ARE OF INTEREST TO YOU

Your resources for choosing them will be the list you compiled through your informal interviews (when Uncle Fred said, "You gotta' talk to my bridge partner... he's got a job exactly like you're describing"), people you already know, and people you have read about or have seen advertised. (One of our most enterprising clients, upon deciding to go into advertising, called the head of every firm listed in the telephone book, getting appointments, leads, and subsequent offers.)

Many career counselors advise that you first write a letter to introduce yourself and explain your purpose, and then follow up with a phone call at a time stated in your letter. (See an example at the end of the chapter.) This approach is businesslike and timesaving in that it circumvents lengthy, sometimes awkward explanations over the phone. Writing a letter to someone you have read about in the newspaper proves very flattering to the recipient. However, you may feel it is more appropriate and comfortable to call directly. A good idea is to experiment with both.

2. ARRANGE APPOINTMENTS BY TELEPHONE

State your name and your referral source. ("I am Julie Job-hunter, and Dr. Smith suggested I should get in touch with you," or "I read the article about your firm in the Tribune.")

State the purpose of your call. ("I am exploring a career change and am very interested in marine biology, but I need to learn more about the field.")

Make it clear that you are not looking for a job but only doing research. ("I am not asking for a job; I am not at that point yet, I would like to hear about what you do, so I can learn more about the field.") You may need to elaborate, since many people won't understand and will think you're asking for a job anyway!

Request a small block of time at their convenience. ("I would appreciate an opportunity to meet with you personally, for about 15 minutes, whenever it's convenient for you.")

3. MAKE EVERY EFFORT TO SCHEDULE PERSONAL MEETINGS

Most people will be willing to see you. We suggest that you begin with people in fields that are not your first choice so you can gain some interview experience with people you don't know. Save the top companies and their executives for future interviews when you have a clearer picture of the field or industry and can ask intelligent, sharply focused questions.

If an individual refuses to meet with you, accept gracefully and ask for a referral. ("I understand how busy you are, Mr. Fish; however, do you think you could recommend another person who might be able to help me?")

4. WRITE DOWN THE AGREED UPON TIME AND PLACE AND THEN RECONFIRM IT

You want to make darned sure you are there when you're supposed to be.

5. PREPARE YOUR QUESTIONS AHEAD OF TIME

We suggest using the following questions as guidelines (you might write them on a card and take it to the interview.):

a. Tell me about what you do. (A good opener. This will allow you to determine if the job requirements match your checklist of criteria.)

b. What do you like about your job? What don't you like? (The latter will be effective in turning up the problems and needs you would face in a similiar position, and also that you will want to address in a hiring interview. People are usually very eager to voice their complaints to a willing listener.)

c. What skills, experience, and training are necessary to enter this field? (as with all questions, don't draw any conclusions based on only one person's opinion. You need several points of view.)

d. What is the earning potential? What are the entry-level salaries? (This is important information to have in salary negotiations.)

e. Who else should I talk to? (Never leave without asking this question. If they can't think of anyone, offer to call back in a few days.)

f. How did you become interested in this field? (People love to tell their life histories. This is a good way to learn about their own career goals and advancement strategies.)

6. TAKE TIME TO DEVELOP RAPPORT

Even though you're prepared with questions, don't interrogate. Relax, make small talk, be casual, and let conversation flow. Ask questions to keep talkative personalities from meandering off into irrelevant storytelling, but above all use the time wisely to create positive relationships while gaining useful information.

7. BEGIN TO TAKE YOUR LEAVE AT THE AGREED-ON TIME

Don't be surprised if you are invited to stay longer, and feel free to do so, but make sure you acknowledge when time is up.

8. LET YOUR INTERVIEWEE KNOW THAT THE MEETING WAS HELPFUL

When you wrap up the conversation, show your appreciation. Letting people know they helped (come on, certainly you can find SOMETHING that was of value) makes them feel important. And any time you make people feel better about themselves, they will feel better about you. ("Thank you so much for taking this time, Mr. Rush. Your suggestion to get into sales for a year is excellent, and I am going to call those two magazine editors this afternoon. I'll let you know what happens.") When interviewees feel they have been of no help, they'll be left with a bad taste in their mouths that they'll associate with you.

9. LEAVE A BUSINESS CARD

10. DO NOT BRING YOUR RESUME TO AN INFORMATIONAL INTERVIEW

Offer to send it if the interviewee asks for one. (Taking your resume to an informational interview says, "I have a hidden agenda." Also sending it later puts you back in their thoughts.)

11. WRITE A THANK-YOU NOTE

Mention again your appreciation, how you benefited, and what steps you have taken, and agree to keep the interviewee posted regarding your progress. Continue to jot notes to them when their leads or suggestions prove fruitful.

12. MAINTAIN AN ORGANIZED FILE SYSTEM OF ALL YOUR MEETINGS

Record the dates, addresses, phone numbers of your contacts; write summaries of the interviews and the items discussed including personal non-job related material. Mr. Rush will be terribly impressed when you meet him again if you ask about his son's soccer or how his bridge lessons have improved his game. Use your contact file cards for this.

By interviewing for information, you are not only meeting and developing rapport with potential employers, discovering hidden job opportunities, and learning about the industry. You are also firmly planting your name in your contact's mind, particularly when you write letters on your letterhead, give business cards, and send thank-you notes. Even if nothing transpires in a particular company, chances are your contacts within it will remember you. Consider Betsy, who decided she was interested in marketing for banks, and hence made an appointment to see the president of every major bank in the city. One chief executive, quite impressed with Betsy's approach to job hunting, happened to be at a party with a gas company VP who mentioned that she needed a customer relations director. The banker, remembering that he had Betsy's card, suggested her for that job, and she got the position. Moral of this story: networking can work for you even when you're absent... and in surprising ways.

IN CONCLUSION

Informational interviewing is a low-stress, enjoyable means of developing profitable contacts and preparing yourself for getting hired. Nevertheless, it requires considerable **patience** (when those phone calls go unanswered and dates keep being changed), **stamina** (when you've scheduled meetings every day, and must constantly remain alert and enthusiastic), and **effort** (making yourself dial that phone one more time; mustering up the courage to call the president of a bank).

Often, you'll find yourself making wrong turns and reaching dead ends. That's an inevitable part of the process. However, if you have defined the kind of work you want and understand how to uncover and investigate possible options, if you are receptive and flexible while remaining focused and directed, and if you're willing to throw yourself wholeheartedly into the search with dogged determination, willing to take the necessary risks and plow on even when you don't feel like it ... you'll find the job you've targeted for yourself, and learn a great deal in the process.

RESEARCH PLANNING CHART

As you begin to identify areas of career interest and specific industries you will want a simple, organized tool for keeping track of potential employers and key contacts within the organization.

The chart below is one method for organizing this sometimes overwhelming phase of job research. Make a separate chart for each area you are exploring and add information as you research and interview for information.

RESEARCH PLANNING CHART

Area of Interest, Position _____

Potential Employers	Contacts in Co./ Position	Address/ Phone	Referral Source

RESEARCHING THE JOB MARKET

Use this worksheet as a guideline for formulating interview questions and recording information about each job you investigate. You'll notice that it will also prepare you for the hiring interview.

Job Title: _____

1. Nature of Work, Job Responsibilities: _____

   _____

   _____

   _____

2. Working Conditions: _____

   _____

   _____

   _____

3. Places of Employment: _____

   _____

   _____

   _____

4. Training, Other Qualifications, and Advancement: _____

   _____

   _____

   _____

5. Employment Outlook (Job Security, Availability) _____

   _____

   _____

   _____

6. Earnings _____

Continues on next page.

7. What interests of mine will be satisfied in this job? _____

_____

_____

_____

8. What personality traits do I have that match those needed for success in this job? _____

_____

_____

_____

9. Which of my needs and values can be met in this job? _____

_____

_____

_____

10. Which skills and special knowledge do I have that are needed for this job? _____

_____

_____

_____

11. Which new skills and special knowledge will I need to develop for this type of job? _____

_____

_____

_____

LETTER REQUESTING AN INFORMATIONAL INTERVIEW

Typing a letter on your business stationery is a recommended method of approaching individuals you do not know.

Correct name and title

Appropriate salutation

State your purpose in writing

Who are you?

Why he should meet with you

Ask for follow-up

Polite closing

Signature

Jeanie Smithson
11200 Alahambra Drive
Kansas City, Missouri 64114

January 3, 1984

Mr. J. B. Connors
Director of Communications
First Union Bank
12th & Baltimore
Kansas City, MO 64106

Dear Mr. Connors,

Jim O'Brien suggested that I contact you to learn more about the communications field.

For five years, I have been employed in personnel management and am ready to make a career switch.

I'm just in the process of exploring and not ready to make a commitment, but I would love to learn more about your field. I would appreciate fifteen minutes of your time at your convenience to ask your advice and get some suggestions.

I will phone you on Wednesday to see if we could arrange a meeting.

Sincerely,

Jeanie Smithson

SUGGESTED READING
Chapter 4

Irish, Richard K., **Go Hire Yourself an Employer.** *New York, Anchor Press/Doubleday, 1973.* The author stresses that good jobs are almost never advertised and one must have inside sources of information to learn of the good openings.

Jackson, Tom, **The Hidden Job Market.** *New York, Times Book Co., 1976.* Another excellent job search guide full of tools, techniques, and exercises for uncovering job leads and organizing the job search.

General Directories and Manuals

The following can be found in most public libraries.

College Placement Annual. Listings of 1500 company and government employers and the kinds of positions they seek to fill.

Dun and Bradstreet's Million Dollar Directory. Two annual volumes listing U.S. businesses and their indicated worth over $500,000.

Moody's Industrial Manual. Annual publication giving facts about a company, a financial statement, products, services and principal executives.

Standard and Poor's Register of Directors and Executives. Published annually with quarterly updates giving information on key people in the corporation – their background and affiliation.

Thomas Register of American Manufacturers. Volumes list American manufacturers by products and services.

State and Local Resources

State Directories of Corporations and Manufacturers. Published by the Chamber of Commerce of the U.S.

Office of the Secretary of State and State Government Library. Information on privately held companies.

Contacts Influential. Listings of local firms and key personnel arranged alphabetically and by nature of business.

Company's Annual Reports. Financial information, directors of the company, and product/service update.

Local Chamber of Commerce. Offers various listings and market information.

Local Chapters of Trade and Professional Associations. Can offer networking, educational seminars, and salary information.

Women's Business Associations. Networking and educational channels.

CHAPTER 5

THE RESUME

It's a basic fact of job hunting: you are going to need a resume.

The best time to start putting one together is NOW, after you have developed a description of the kind of job you are seeking.

PURPOSE

Why is a resume so important? First of all, you don't want to be caught without one when you are asked for it. And it is likely you will be asked often, whether it be by your next door neighbor, the president of Widgets, Inc., or in response to a want ad. A resume is an important part of the networking process. People will want your resume for whatever reason: "In case something comes up, let me keep it on file," or "If I hear of anything, I'll pass it along," or "Send in your resume, and we'll take a look at it." Resumes are "memory joggers" to be sent to individuals after you meet them in informational interviews. They also serve as introductions to get you in the door for job interviews.

In addition, the very process of putting together this document helps you organize your thoughts for the hiring part of your job search. Thus, we encourage you to write the resume yourself. Having to think through your skills, abilities, and experience enables you to be more articulate in the job interview. It may be laborious, but well worth the effort in the long run.

THE PROCEDURE

We know that the task of writing a resume can be tedious and time consuming. We strongly recommend that you do not try to organize and write the resume all in one sitting. Set up a schedule such as the following:

1. **Begin by putting on paper all your past jobs.** (paid or volunteer) listing everything you did, achieved, accomplished and enjoyed in each one. (See Resume Inventory Sheets at the end of this chapter.)

2. **Take a break for a few hours.**

3. **Scan resume writing books,** looking at examples of resumes.

4. **Spend 1 - 2 hours organizing the information** you first wrote out.

5. **Take a break.**

6. **Refine,** modify, edit, organize.

7. **Take a break.**

8. **Again, refine, and edit.**

9. **Continue taking breaks,** then polishing each draft. This goes on for at least a month. The rhythm of backing away then rewriting keeps you fresh, alert, and creative.

10. **Ask friends to critique** your draft.

FORMAT

You have a choice of three formats and several alternatives. Information should be organized under broad headings. The headings depend on what format you choose. You can be creative, however, devising variations of what we suggest. Whatever format you choose, we urge you to include a job objective. Sample formats are included at the end of this chapter.

CHRONOLOGICAL

The chronological resume is a listing of your work experience and educational history in chronological order. The major categories are generally **Work History, Education, and perhaps Honor Achievements and Personal Data.** All categories highlight previous job experience and education, with short, brief descriptions. This format is most commonly used by someone changing jobs in the same field where continuous work activity in that field needs to be documented.

FUNCTIONAL

The functional resume organizes your experience according to skills or functions. The major headings are **Skill Categories,** and the information under each skill demonstrates your capabilities through a variety of work, volunteer and personal experiences. This format is advisable for a career changer, a volunteer or a person who wants to re-enter the work force after a period of absence.

COMBINATION

Categories are a combination of **Skills and Chronological Work History.** This format affords both the job changer and career changer an excellent opportunity to present capabilities and transferable skills combined with a record of work experience. We generally encourage this type of resume.

ALTERNATIVES TO THE RESUME

The Qualifications Brief and Biographical Sketch:

This can be used by a person who is self-employed and needs to market their services in terms of their own qualifications and professional expertise. For example, a professional consultant whose objective is to market her speaking, facilitation, and consulting skills would create an information sheet describing her education and professional qualifications. This would be documented with specific mention of companies and organizations that she has worked with as well as a description of the types of presentations and services she could offer. These may be written in paragraph form rather than the one-liners found in resumes.

The Career Portfolio:

This is a cumulative record of your accomplishments as a volunteer, a student or professional. It includes examples of creative work such as paintings, writing samples, or designs and layout which are related to the position for which you are applying. Include only a few examples of your best work. An overflowing portfolio overwhelms, detracts from your verbal presentation, and diminishes the importance of your work. For a volunteer, this type of portfolio could be used in interviews for other volunteer or paid positions because it helps to document the valuable skills gained in a volunteer setting. It might include recommendations, examples of public recognition, and samples of written work.

WRITING THE COVER LETTER

A resume without a cover letter is like an unannounced salesperson showing up at your door. If you are going to let in a perfect stranger, you at least want to see their credentials. This is exactly what a cover letter does – it introduces you, a total stranger, to the reader. It must be compelling, personable, and brief. It needs to specifically relate to the position in question. Remember, you only have eight seconds to convince the reader to invite you in. The three parts of the cover letter are:

The Opening:

You can begin with a statement about your purpose in sending your resume. Why are you writing? What do you want from them? Are you responding to a want-ad or a tip on a job opening from a friend. For example, "Jane Smith, the editor of Consumer Guide, suggested that I contact you about a position as managing editor of your journal."

The Main Body:

In 2 or 3 sentences, describe who you are and identify your unique strengths and abilities that could benefit the employer. Emphasize and support these. Your research of the company can be helpful here in addressing specific needs of the company. For example, "I hold a journalism degree and have five years of experience as an editor for technical journals. For the last two years, I have supervised three other assistant editors."

The Closing:

Express your willingness to provide more information by letter or in a person. State a more specific time to recontact them or more generally say, "I'll be available to meet with you at your earliest convenience."

DO'S AND DON'T OF RESUMES AND COVER LETTERS

Consider the following hints for effective resume and letter writing. We strongly urge you to review samples of resumes in this chapter as well as in the books that have been listed as good resources. Remember to be creative and professional.

DO:

1. Use plain English; not jargon.
2. State your job objective. It is difficult for an interviewer to know your direction and career goals unless you tell him or her.
3. Watch your spelling and punctuation. Avoid typographical errors. (First impressions are the deciding factor.)
4. Use factual, concise language; include action verbs and adjectives to help describe your skills and accomplishments.
5. Emphasize your positive skills and abilities, but be truthful. Some of the information will be verified.

6. Use good, bonded paper similar to your business stationery for printing your resume. Typesetting or the use of a word processor is recommended.

7. Have at least three people critique your resume before you draft your final copy.

8. Keep old resumes for future use.

9. Send a cover letter with each resume. Type the letter on your business stationery.

10. Use numbers and figures to show results. Not "I managed a staff," but "I managed a staff of 15." Not "successfully marketed new products" but "increased sales of new products 31% by developing creative marketing strategy."

DON'T:

1. Send carbon copies.

2. Include salary information (it's negotiable, so why limit yourself?)

3. List references on your resume. Prepare a typed list of references and include: name, position, address and phone. Take this with you to the interview.

4. Address a cover letter to "To Whom It May Concern" - always have a specific name.

5. Include personal data, such as weight, age, and marital status. You never know when this can work against you.

To help organize and write your resume, we have included worksheets, examples, and a suggested reading list.

Exercise One

RESUME INVENTORY SHEETS

Fill out a sheet for every job you have ever had, including voluntary positions, for the past ten years. It may help to go back to the skill lists in chapter two to give you a vocabulary of skills. This becomes the data bank for your resume.

JOB TITLE: _____
(eg. Personnel Specialist)

ORGANIZATION: _____   **DATES:** _____
(eg. Community Hospital)   (eg. 1978 & 1981)

What Did You Do? Specific Activities	What Skills and Special Knowledge Did You Use?	What Was A Positive Result?

Exercise Two

WRITING SKILL PHRASES

The purpose of this exercise is to help you write statements that demonstrate your skills. Therefore, complete the steps below.

1. Review your Skill Clustering Chart from chapter two.

2. Add any new skills from your Resume Inventory Sheets.

3. Choose the top 2 - 3 skill clusters that include the majority of your skills.

4. Write a short descriptive phrase telling how you can use each skill. Use adjectives, action verbs, and nouns to specifically illustrate each skill.

5. Rewrite and rewrite. Do not try to make your first attempts perfect.

EXAMPLE:

Skill Cluster No. 1   *ADMINISTRATION*

Direct a staff to achieve organizational objectives.
Develop and administer departmental budget.
Serve on administrative team as advisor.
Interview, hire and terminate employees.
Present program proposal to management.

Skill Cluster No. 2   *PROMOTIONAL*

Communicate concepts, policies and procedures to all levels of management and staff.
Promote organizational image and programs to community.
Build and maintain contacts.
Conduct market research studies and recommend. alternatives.
Design marketing strategies for implementing new services in the community and business.

Skill Cluster No. 3   *PLANNING AND ORGANIZATION*

Plan and coordinate complex activities for multi-unit organization.
Develop effective management and delivery systems.
Develop hiring policies.
Design and implement training programs.

See exercise on next page

Skill Cluster No. 1

_____

_____

_____

_____

Skill Cluster No. 2

_____

_____

_____

_____

Skill Cluster No. 3

_____

_____

_____

_____

Exercise Three

ORGANIZING THE RESUME DATA

Begin writing a rough draft which compiles the information gathered in the previous two exercises.

It is helpful to think of the resume as being organized into three parts.

I. Personal Information (name, address, telephone numbers)

II. Job Objective (kind of position, using these skills, in this kind of organization, solving these problems)

III. Past Experience (skill offerings, work experience, education, honors, achievements)

Below, we have given you an illustration of how this information may be presented. The left-hand column is a description of the resume category. On the right are specific examples this person chose to use.

Jim Potts
3800 W. Armourdale
Topeka, Kansas 66609
913-267-5896

Job objective describing the kind of job you are seeking

Job Objective:
Management position which integrates my administrative, planning and organizational skills to promote new services and programs.

1. Major skill cluster with skill phrases describing each sub skill or 1 list of skills.
2. Prioritize all skills.
3. Use present tense of an action verb.

Capabilities:
Administration
- Direct a staff to achieve objectives
- Communicate policies and procedures to staff

Planning and organization
- Develop effective management system
- Plan and coordinate complex activities for a multi-unit organization

OR

Capabilities:
- Plan and coordinate complex activities for a multi-unit

66

organization
- Develop effective management system
- Direct a staff to achieve objectives
- Communicate policies and procedures

Dates of Employment (most recent first), position, company/organization and location

Work Experience:
1979 - Present
    Director of Personnel
    Community Hospital
    Topeka, Kansas

Brief description of duties

Managed 500 plus employees professional and hourly Responsible for all human resource functions

Cite accomplishments that support capabilities (past tense of action verbs)

- Served as member of policy-making administrative staff of a $45 million business
- Designed and implemented first employee training program

Degree, institution, date, honors, activities

Education:
1980, MBA, Management
University of Colorado
1976, B.A. Personnel
    Administration
University of Colorado

Affiliations, honors, professional responsibilities and activities that capabilities support

Professional:
Personnel Management Association: Chairman of Community Relations, responsible for promoting the professional image of the organization and the credibility of its members to the business community.

Chamber of Commerce: Finance Committee, worked with small businesses to create more economic activity and development of long range plans.

Community activities

Community:
United Way Chairperson, supervisor of 30 person fund-raising team.

Language, fluency, military experience, extensive travel, and anything that is pertinent to job objective

Personal:
Native of the Topeka area for twenty years

Type references on separate sheet of stationery for interview purposes

References:
Available upon request

CHRONOLOGICAL

Recent college graduate. Emphasize education and extracurricular activities.

Steve James
127 E. Seventh
Lawrence, Kansas
(913) 751-2292

Job Objective: Seeking a position with advancement opportunity that combines a mathematics background with business administration.

Education: University of Kansas, Lawrence, Kansas

B.A. Mathematics, 1981

Special emphasis in business administration, including computer programming, accounting, economics and management.

Extracurricular Activities Treasurer for national social fraternity—responsible for all financial transactions and obligations and for collecting payments from members and maintaining files and records.

Vice-president for senior class—responsible for public relations within the university and with the community.

Experience:

Summers 1979 & 1980 Macy's, Kansas City, Missouri

Sales clerk in men's and boy's department
• Increased experience and responsibility with sales, customer relations and inventory work;
• Asked to work in other departments because of adaptability and communications skills.

Summer 1978 Macy's, Credit Department, Kansas City, Missouri

• Received and filed payments;
• Researched applications for credit;
• Established consumer relations with customers.

Summer 1976 McDonalds, Overland Park, Kansas

• Cashiered and did bookkeeping for the manager.

References: Available upon request

CHRONOLOGICAL

Chronological Format for Changing Jobs in the Same Field. Continuous work experience is best documentation for changing jobs within the same field.

```
                                    Steven H. Green
                                    1201 W. 31st Street
                                    Washington, D.C.   20009
                                    (202) 367-2018


Job Objectives:         Seeking a position as a sales representative in the
                        computer technology industry with opportunity for
                        advancement.

Work Experience:

  1980 - 1982           Sales Representative
                        Textron Corporation
                        Bethesda, Maryland

                        ● Conducted market research for two-state region
                        ● Covered Maryland and Virginia as sales representative
                          with annual sales volume of 1 million dollars
                        ● Served as sales trainer for new representatives

  1979 - 1980           Sales Representative Trainee
                        Data Systems
                        Baltimore, Maryland

                        ● Trained in data processing equipment sales
                        ● Opened 20 new accounts within first nine months as
                          sales representative

  1977 - 1979           Assistant Manager
                        Radio Shack
                        Washington, D.C.

                        ● Sold merchandise over quota for 6 consecutive months
                        ● Managed store's personnel


Professional:           United Computer Systems Association


Education:

  1973 - 1975           Community College of Washington, D.C.
                        Emphasis on business and computer technology


References:             Available upon request
```

FUNCTIONAL

Re-entry with emphasis on volunteer work experience. Functional Format for de-emphasizing lack of work experience.

Glenda Harrison
125 Euclid
Tiburon, California
(415) 435-7187

Job Objective: Seeking a position in promoting fundraising for an organization or institution

Capabilities:

Promote Fundraising:
- Organize and implement promotional campaigns
- Solicit contributions from community and business
- Sell ideas, products and services

Manage:
- Plan and coordinate details for a campaign
- Manage and supervise staff and volunteers

Communicate:
- Write press releases and promotional materials
- Write proposals for state funding
- Serve as a liaison between organization and community

Work Experience:

American Diabetes Association
Coordinator of Fundraising Campaign
- Raised over $200,000 for Diabetes Association through telethon

Voluntary Action Center
Chairman of Public Relations Committee
- Created innovative programs for promoting services in community which resulted in doubling the number of people who used Center's referral services

Education:

1971 University of Southern California
B.A., Sociology

References: Available upon request

COMBINATION
FUNCTIONAL/CHRONOLOGICAL

Emphasize skills and work accomplishments.

Jim Potts
3800 W. Armourdale
Topeka, Kansas 66609
913-267-5896

OBJECTIVE

Management position allowing me to utilize my planning, organizing, and administrative skills to promote a product or service.

CAPABILITIES

- Plan and coordinate complex activities for multi-unit organization
- Communicate concepts, policies, and procedures with all levels of management and staff
- Direct a staff to achieve organizational objectives
- Conduct comprehensive research studies
- Analyze research data, consider alternatives, and recommend effective solutions
- Write and edit professional reports, policies, and procedures

WORK EXPERIENCE AND ACCOMPLISHMENTS

Director of Personnel, Community Hospital, Topeka, Kansas
March, 1979 to present
Responsible for all aspects of the human resource management function for 500+ professional and hourly employees providing health care services.
- Served as a member of the policy making administrative staff of a $45 million business
- Researched and wrote new employee handbook and coordinated a committee in the preparation of a supervisory manual
- Updated comprehensive benefits program resulting in cost savings and improved services to employees
- Advised management personnel in all areas of manpower planning and utilization

Director of Support Services, Community Hospital, March 1978 - March 1979

Responsible for staffing, budgeting, and development of policies and procedures for a multi-unit division providing clinical and support services.

- Developed and administered $6 million budget for a multi-unit division consisting of eight departments and 120 employees
- Conducted a manpower utilization study with an outside consulting group which resulted in increased productivity and a $62,000 salary savings
- Supervised a staff of professionals providing key support services including materials management, records administration, and food service

<u>Personnel Specialist</u>, State of Colorado Office for Planning and Programming, February 1976 to March 1978

Provided consulting services to county and municipal governments in establishing and/or refining personnel management systems. Served as the personnel officer for this state agency.

- Developed total personnel management systems for counties and cities which previously had no formalized systems
- Provided a variety of technical services to governmental units including designing salary systems and classification plans and advised concerning FLSA, EEOC, collective bargaining, and preventive labor relations.
- Presented program proposals to top governing boards in a manner which encouraged acceptance and implementation

COMMUNITY ACTIVITIES

United Way Chairperson, supervised team of 30 fund-raisers.
Board of Deacons, Valley Presbyterian Church

EDUCATION

1980, MBA Management University of Colorado
1976, BA Personnel Administration, University of Colorado

RESUME ALTERNATIVE: The Biographical Sketch

JANE M. SOUSA

As a career consultant, Jane Sousa is in the business of conserving human resources. She helps individuals find satisfying career directions and helps employers stimulate productivity using an innovative career planning process.

Ms. Sousa helps individuals identify their most valuable resources - their skills and abilities and turn them into a set of criteria for selecting the right career, the right job, and in the employer's case, for increasing productivity.

In addition to counseling individual job hunters and career changers, Ms. Sousa has designed and conducted numerous seminars and workshops for such organizations as Women in Communication, area College Marketing Associations, Women in Radio-Television, Home Economists in Business, Missouri Press Club and American Management Association. She has also lectured at the University of Missouri-Kansas City, and has participated in career seminars sponsored by Penn Valley and Pioneer Community Colleges and the federal government.

Workshop and seminar topics for special interest groups and companies include . .

- Skills Analysis
- Success Strategies
- Self-Marketing Techniques
- Researching the Hidden Job Market
- Organizing the Job Search
- Goal Setting
- Resume Writing
- Interviewing Techniques
- Job Burnout/Job Enrichment
- Time/Stress Management
- Leadership Development
- Office Politics

Ms. Sousa has served as a resource for newspaper articles and has appeared on local television and radio programs.

A graduate of the University of Missouri, Ms. Sousa spent seven years as a secondary teacher in the Kansas City, Missouri School District. She holds a M.A. in Guidance and Counseling from the University of Missouri-Kansas City and is a member of the American Personnel and Guidance Association, the National Vocational Guidance Association, and the American Society for Training and Development.

COVER LETTER FOR A RESUME

Steve Green
1201 W. 31 Street
Washington, D.C. 20009
January 10, 1984

Mr. William Needle, President
Robot Technology, Inc.
217 W. Seventh Street
Chicago, Illinois 60062

**Purpose
in
writing**

Dear Mr. Needle,

Recently I learned through Dr. Eugene Mott, a professor in computer
science at the Community College of Washington, D.C., of the expansion
of your company's sales operations to Maryland and Virginia. I would
like to be considered for any newly created sales positions.

**Address your
qualifications
to those
given
in ad**

I have a broad knowledge of computer technology, having completed course
work in the field as well as having work experience in computer sales,
promotion and market research. As a sales representative with a computer
hardware firm, I was responsible for a two-state region which brought in
sales of 1 million dollars in one year.

For your review, I am enclosing my resume. I would appreciate a personal
interview with you to learn more about your company and to discuss my
application. If you plan to be in Washington in the near future, please
let me know and I can arrange to meet with you.

**Request
follow-
up**

Sincerely,

Steve Green

enc.

LETTER ANSWERING A BLIND ADVERTISEMENT

Susan Schmidt
6801 N.W. Oak
St. Louis, Missouri 63141

Mr. James Tobin
P.O. Box 2301
Kansas City, Missouri 64110

Dear Mr. Tobin,

Purpose in writing

In response to the advertisement in the Kansas City Star, December 27, I am making application for the position described as an experienced writer for an advertising department of an industrial corporation.

State qualifications

My qualifications as detailed in my resume, match the described position. I have had an internship in industrial advertising as well as a B.A. in English with additional course work in the area of writing and communications skills. My writing experience has included writing and editing brochures, sales promotions, press releases, and magazine articles. My background had familiarized me with the consumer population to which you are directed, and which I feel greatly enhances my qualifications as a communicator.

Ask for follow-up

I am available for an interview and would like the opportunity to discuss this position and my qualifications in more detail.

Sincerely,

Susan Schmidt

enc. resume

THANK YOU LETTER FOLLOWING AN INTERVIEW

Jim Baker
135 South Past
San Francisco, California 94108

January 1, 1984

January 1, 1984

Ms. Carla Overton
Hallmark Cards, Inc.
P.O. Box 1234
Kansas City, Missouri 64108

Expressing your positive impressions

Dear Ms. Overton,

    I greatly enjoyed our discussion today regarding the position of new product manager at Hallmark Cards.

Re-state qualifications

    I was very impressed with the plant operation and the personnel. I strongly believe that my management experience and technical abilities qualify me for this position with Hallmark Cards, and appreciate your consideration of me for this job.

Closing

    Thank you for your time and interest. I look forward to hearing from you soon.

            Sincerely,

            Jim Baker

SUGGESTED READING
Chapter 5

Jackson, Tom, *The Perfect Resume. New York, Anchor/Double-day, 1981.* How to construct a resume that will be better than 98 percent of those in the market today. Many sample resumes for different occupations.

Lathrop, Richard, *Who's Hiring Who? Berkeley, Ten Speed Press, 1977.* Rather than a resume, he recommends a qualification brief. He offers excellent examples. The best on the market in our opinion.

Resume Prep Manual for Women. Catalyst Publications, 14 East 60th Street, New York, New York, 10022.

66 *When you get into a tight place
and everything goes against you,
till it seems as though you could
not hold on a minute longer,
NEVER give up then, for that
is the place and time the tide
will turn."*

— Harriet Beecher Stowe

CHAPTER 6

GETTING HIRED

*An interview is to employment what courtship
is to marriage.*

Eventually in your job search you will find that you have col-
lected enough information to be sure of your direction. The change
will come when someone expresses an interest in you or you have
pinpointed an organization you'd like to work for or a job you'd like
to land. At this point, you will actively seek to set up a new kind of
appointment: the *HIRING INTERVIEW.*

MOVING TOWARD AN INTERVIEW

The **traditional routes** to the hiring interview include the
following:

- **Responding to want-ads**. A good idea. Sometimes jobs
  result, but even when they don't you will get a lot of interviewing practice.

- **Mailing out unsolicited resumes.** Exhausting and by-
  and-large unproductive, since companies respond to about
  one out of every 245 resumes received.

- **Contacting a company's personnel department.** Not
  usually very fruitful, since the main function of these departments is to screen applicants.

- **Visiting employment agencies.** Watch out for reliable
  ones.

Here's **a more creative approach** to setting up a hiring
interview:

- **Call your information contacts** in your targeted companies and let them know you have finished exploring
  possibilities and have come to a decision.

- **Let them know that you benefited from their help** and
  were impressed with their organization.

- **Tell them you would appreciate any leads** and would like
  the opportunity to talk about what you could do for them as
  an employee of their firm.

- **Contact people you haven't met but who have possible job openings** to discuss your career exploration and their organizations. This will get you in the door of a person with the power to hire and helps you to continue the research and decision-making process.

PROCEDURE

Do beads of perspiration form on your forehead and trickle down your armpits at the mere mention of a job interview? Relax. There's no need for panic. We can show you how to control the interview and turn it into a positive experience. How? Let's put this formidable event into perspective. Recognize that you are selling a product–yourself–and that you have very limited time (26 minutes average), in which to convince a potential employer that you are the best for the job. There are four important steps to achieving this goal:

1. **Prepare ahead of time.**
2. **Concentrate on building rapport.**
3. **Be specific in explaining how you can benefit the company.**
4. **Follow the prepared guidelines enumerated later in this chapter.**

Let's elaborate each of these points.

1. PREPARE, PREPARE, PREPARE

If you have followed the steps in the preceding sections, you have done a lot of homework already. You already know yourself, and you are clear about the position you are seeking, why you want it, and what personal qualities and skills will enable you to perform it well. Interviewers generally have no patience for wishy-washy applicants with fuzzy goals and poor self-knowledge.

Once you have targeted a company, make yourself familiar with both the organization and the field. If your informational interviewing yielded only an overview, take steps to find out more specific information:

- *Ask a stockbroker for literature on the company.*
- *Ask the reference librarian at a local library for ideas and information.*
- *Talk to employees (past and present) and to employees of competitors.*

- *Take tours of the facilities (and of the competition's).*
- *Pay attention to media ads and articles on the company.*

In pursuing these steps, find out all you can about:

- *The company's problems, challenges, needs and goals.*
- *Its philosophy and policies.*
- *Its products and services.*
- *Its age, size, growth rate and growth pattern.*
- *Its future projections, goals, and values.*
- *The competitors of the department in which your job would be.*
- *The names of key individuals in the firm.*

Equipping yourself with such information for a job interview will both reinforce your self-confidence and impress the interviewer with your interest in the firm.

Do yourself a favor: make an individual "prep sheet" for each interview. Record every piece of information and every idea that could possibly be useful in the interview itself. A potential employer will be impressed by your initiative.

2. CONCENTRATE ON BUILDING RAPPORT

An interview is to employment what courtship is to marriage: an opportunity to make an impression and build a relationship that will lead to something more permanent.

In concentrating on selling yourself and trying to impress another, it is easy to forget that hiring decisions are usually arbitrary and subjective. The *Wall Street Journal* reported a study that found that more hiring decisions are based on personal chemistry than any single factor, including skills.

But how do you create the right chemistry? Realize that it won't always happen. Just as not every date ends with a marriage proposal, not every interview concludes with a new job offer (thank goodness for both!). Nevertheless, you can maximize your chances by using a little psychology.

All human behavior is motivated by needs and the drive to satisfy them. Therefore, your trump card lies in identifying another's needs and showing how you can satisfy them.

In the context of the hiring interview, two kinds of needs are of particular interest. The first are emotional needs. Your interviewer, being a mere mortal, possesses two very basic needs (cravings or hungers, some psychologists call them): the need to feel important and the need to be loved. You can fulfill both in a job interview. The trick, above all else, is to concentrate on making the interviewer feel good about him or herself. When you express admiration and exhibit warmth and enthusiasm, people will respond favorably to you. They will feel that you like them and consequently they'll feel good about themselves. They'll relax and enjoy the encounter. Threaten another's self-esteem, and his or her defenses are bound to shoot up. The relationship crumbles, and you've lost your chance.

In preparing for the interview, you should have identified the specific needs of the firm, department, position, and boss. Sell yourself by demonstrating how your skills and experience can satisfy those needs. Consider Jack, who noticed that store sales were declining and who, in a hiring interview, suggested a promotional campaign that included special events for the customers in the store while they shopped. Jack knew that his skills and expertise lay in creative promotion, and he was able to convince the store managers that he could solve their problems.

3. SELL BENEFITS

The information you have been so diligently gathering is of little value if not astutely applied. Don't simply tell your interviewer what you are capable of doing or even how you can fill their needs. Spell out the benefits they will receive by hiring you.

Remember, everyone operates from a position of self-interest. Interviewers are more likely to want you when they understand how you can benefit them directly and specifically.

4. FOLLOW THE GUIDELINES

Use the following guidelines as a checklist before, during, and after the hiring interview.

Before the Interview:

- *When you set up the appointment, be sure to write down the name, address, and time, and then verify this information before hanging up.*

- *Show up a few minutes early.*

- *Dress conservatively, avoiding extremes. While your appearance won't get you the job, it can definitely disqualify you.*

Bring the following:

- *A note pad and pencil in case you need to record information during the discussion or make notes afterwards.*

- *At least two copies of your resume (even if the interviewer already has one on file).*
- *Your Social Security number.*
- *Names and addresses of at least three references typed on your stationery and any letters of recommendation you might have.*
- *A portfolio of your work with samples and any supportive materials (such as articles about yourself).*

At the opening of the interview:

- *When introduced, make a conscious effort to remember names (better yet, try to learn them in advance).*
- *Greet the interviewer, making sure to pronounce his or her name correctly.*
- *Let the interviewer take the lead in shaking hands and sit when asked.*
- *Avoid smoking or chewing gum, even if invited to do so.*
- *Engage in a few minutes of small talk to break the ice and establish rapport. Look for common interests ("I see by the picture on your desk, your son plays soccer...so does mine").*
- *Body language is a powerful tool: lean forward, make eye contact, and look interested, but don't exaggerate.*
- *Have an opening statement prepared (you may not use it, but you'll feel better just having something to say).*

In getting down to the business of the interview:

- *Keep your responses to the point. The interviewer wants to find out as much relevant information as possible in a limited amount of time. In essence, he or she wants to know (1) are you competent? and (2) will you and the firm be compatible?*
- *Think before you talk.*

- *Talk in specifics, not generalities. (For example, rather than "I like to work with people," say "My experience in the Peace Corps has taught me how to get along and motivate people under all kinds of conditions.")*

- *Don't mumble or ramble on.*

- *Describe your accomplishments, not your duties. (Rather than "I was supervisor," say "I supervised thirteen people and had responsibility for a budget in excess of $750,000.")*

- *Be sensitive to the interviewer's body language. When people's eyes stop focusing on you, and they start shuffling papers, they are sending you a message, such as "I'm losing interest" or "I don't follow." Respond to the silent message by altering your course–for example, say "Perhaps I should say this in another way" or "Am I making myself clear?"*

- *Be prepared with the major points you want to make and possible answers to tricky questions. Ask yourself, "If I were in the interviewer's seat, what would I want to know?" (Practice on a tape recorder or with a partner, making sure you create opportunities to inject these points.)*

- *Highlight the parts of your experience and the skills that directly relate to the position.*

- *If the interviewer tests your mettle with intimidating or stressful techniques, keep cool, be diplomatic, and remain gracefully assertive. Don't lash back or attack.*

- *Welcome objections. Remember that as long as they are unspoken, they will be barriers to your getting ahead. Get them out and overcome them. Then, no matter how chauvinistic or illegal the objections are, if you want the job, pleasantly and firmly prove them to be unfounded.*

- *Always be honest. Even though you need not volunteer negative information, answer truthfully when questioned.*

At the close of the interview:

- *Let the interviewer initiate the close.*

- *If you are offered a job, you need not respond on the spot. Take some time to think it over.*
- *No job offer? Don't be discouraged. This is normal for the first few interviews, since all candidates must be carefully considered. The purpose of the first, even second, interview is to get invited back for another. The interviewing process can be a long, drawn-out affair.*
- *Don't talk money until you are offered the job. You will be in a better negotiating position that way. Talk opportunity, not security, in the early stages.*
- *Make sure you leave on a positive note. Shake hands, indicate your interest, and ask when a decision will be made.*

The follow up:

- *Evaluate your experience, but don't punish yourself unmercifully for anything less than perfection. Did you connect and get your points across?*
- *A well-written follow-up note is appropriate. It should convey an image of confidence, competence, and enthusiasm. Include or reiterate important points.*
- *Be persuasive, not pesty, in your follow-up calls.*
- *If you get the job, congratulations.*
- *If not, use the experience as a lesson for your next interview.*

We encourage you to contact the interviewer and ask for feedback in order to evaluate your interview performance. Although asking for such information can be awkward, such a discussion can be invaluable, even yielding other job opportunities and leads.

Above all, don't be discouraged or defeated when interviews don't pan out. The path to becoming happily employed is paved by no, no, no, no, but it leads inevitably to a final **yes.**

EMPLOYMENT INTERVIEW CHECKLIST

Complete this sheet prior to each interview you have. It will help you to organize your presentation and build your confidence for the interview.

EMPLOYER INFORMATION

(based on your research and networking activities)

Position being interviewed for: _____

Employer _____ _____ Department _____

Interview location _____
              Street   City   State   Zip

Phone _____ Date/Time of Interview _____

Contact person(s) _____

Description of position being applied for _____

_____

_____

Salary range _____

Products/Services of Employer _____

_____

Customer Profile _____

_____

Competition _____

_____

Public Image _____

Recognized Problem Areas _____

_____

Future Company Directions _____

_____

_____

ABOUT YOURSELF
(based on your self-assessment and focusing exercises)

Skills and Abilities _____

_____

Work/Personal Accomplishments _____

_____

Job Criteria (prioritize) _____

_____

Anticipated Problem Areas _____

_____

Questions to ask the Interviewer _____

_____

Other Information _____

_____

Practice Interview Questions

Take some time to develop a response to these questions before you first interview. Enlist the help of a friend.

What do you look for in a job?

What are your career goals?

How good is your health?

Can you work under pressure, deadlines?

If you began again, what would you do differently?

How do you rate yourself as as a professional? as an executive?

How would you describe the "ideal boss"?

Do you prefer staff or line work? Why?

Are you a leader? Give an example.

Are you creative? Give an example.

What interests you most about the position we have? the least?

What features of your previous jobs have you disliked?

What one thing would you like to accomplish?

Why are you leaving your present position?

What is your philosophy of management?

Why should we hire you?

What can you do for us that others cannot?

How long would it take you to make a contribution to our firm?

What position do you expect to have in five years?

What other types of jobs are you considering? What companies?

What kind of salary are you worth?

How would you describe your own personality.

Would you describe a few situations in which your work was criticized?

What is your biggest strength? Weakness?

What would you like to learn?

SUGGESTED READING
Chapter 6

Bostwick, Burdette E., *111 Techniques and Strategies for Getting the Job Interview.* New York, John Wiley and Sons, 1981. A comprehensive approach to the subject of interviewing with discussion of ways to get interviews, how to conduct your own P R campaign and the interview itself.

Cogger, John and Morgan, Henry, *Interviewer's Manual.* New York, Psycological Corporation, 757 3rd Ave., N.Y., N.Y. 10017. Valuable tips on preparing for and handling the job interview.

Irish, Richard K., *Go Hire Yourself an Employer.* New York, Anchor Books, 1978. Tells you how to find an employer and effectively interview the interviewer.

Medley, H. Anthony, *Sweaty Palms: The Neglected Art of Being Interviewed.* California, Life Time Learning, 1978. Easy to read and chock full of clever tips for taking the fear out of interviewing.

National Business Employment Weekly. Wall Street Journal. An excellent publication for uncovering job openings across the country.

66 *You don't get paid for effort. You get paid for results."*

— Derek Newton

CHAPTER 7

SALARY NEGOTIATION

Once you receive a job offer, it's time to talk money.

For whatever reason – fear of rejection, lack of confidence, a wish to avoid conflict, or early conditioning – many people often shy away from negotiating about salary, hastily agreeing to whatever is offered or accepting the first "no" as final. A seemingly endless number of books and articles have been written to describe the important, often complex process of salary negotiation, and this attention to the subject is well-warranted.

This chapter covers the main points of negotiation strategy and some techniques to help you in your hiring interview. However, considering the breadth of the topic, we strongly suggest that you look through some of the recommended books we have listed.

THE PROCESS

Let's start with a basic definition of **negotiation.** It is a give-and-take communication process that allows all parties with vested interests in the outcome to reach a mutually satisfying agreement. Underline those last three words. The purpose of negotiation is to arrive at a point where all parties involved feel good about the outcome. Everyone wins something. Be aware of win-lose situations, for losers can get angry and nasty and want revenge. You may not get everything you want, but if you approach the task thoroughly, you'll be prepared and mentally primed, and you'll surely come out a winner.

THE REQUIREMENTS

What goes into your head before you open your mouth is the foundation for a successful negotiation. In other words, your success will depend on thorough preparation. If you have followed the previous steps outlined in this book, you have already laid much of the groundwork.

Preparation for salary negotiation should include the following:

Knowledge of the going rates in your field. Familiarize

yourself with salary ranges by asking direct questions during your informational interviews. In addition, talk to college placement centers, study want-ads, call employment agencies, contact trade and professional associations, visit the public library, and read annual reports and company literature.

Understand your bargaining power. A job offer automatically invests you with bargaining power. You've been offered the job, so obviously you have something the firm wants. To assess your worth, analyze the competition, identify your special skills, and take a hard look at what you bring to the relationship. Then, try to identify the employer's objectives and to assess the pressure they are under to fill the position.

Plan what you are going to say and how you will say it. Identify the major points you want to make. You want to summarize the requirements of the job and match your skills and abilities to them. Write them down if you like. Develop convincing arguments to possible objections. Remember, the employer *expects* to bargain. Your task is to convince him or her that what you bring to the relationship is valuable and should be reimbursed accordingly.

Practice ahead of time. Find a tape recorder–or better yet, a sparring partner to play devil's advocate and force you to think through answers to difficult questions. Also, mentally rehearse a successful encounter. Vividly imagine yourself achieving your goals.

Work on shaping a positive attitude. While you diligently prepare, give yourself some pep talks. Attitude is a crucial determinant to success. Research shows that the most successful bargainers are those with the highest aspirations, who make the highest demands, and have the greatest respect for the other side. Remind yourself again and again that the firm would never have offered you the job if you didn't have what they wanted. Remember, too, that people tend to put a greater value on what they pay a high price for.

If the salary and benefits you are asking for are realistic, approach the negotiation with a cooperative, friendly, but persistent attitude. Let the employer know that you are willing to listen and evaluate all the options in order to find a solution that is acceptable to both of you. Refuse to settle for less than you deserve.

If you are afraid that asking for more than the original offer will cause the firm to retract the offer, if you fear that never again will such an opportunity come your way, or if you see negative responses as personal rejections, then you will be unable to negotiate effectively. You'll throw your chips away and come from a position of weakness rather than strength. If you choose to do this, realize that selling yourself short can eventually lead to frustration and dissatisfaction with the job.

A healthier attitude is believing that you have the power to make an impact and that you have everything to gain from the experience (if nothing else, you gain the practice you need to perfect your negotiation skills).

PROCEDURE

Below are the most important factors in successful negotiating:

1. **Timing is very important.**
   Discuss salary only after you have been offered the job. In the course of the interview, should you be quizzed about salary requirements, you might respond, "I can't answer that without knowing more about the job."

2. **Don't underestimate the power of your appearance.**
   How you look and how you express yourself weigh heavily on the outcome. If you're seeking a professional position and comparable salary, dress businesslike.

3. **Let the interviewer make the first offer.**
   If asked what you feel you deserve, toss the ball back by responding, "What range did you have in mind?"

4. **Think in terms of salary ranges.**
   Identify a realistic figure and then ask for more so you have room to bargain. Career specialist Richard Bolles advises a range that "hooks" onto the one given or estimated by you in this way:

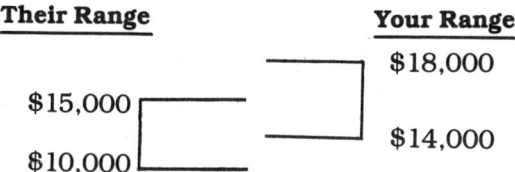

Their Range	Your Range
	$18,000
$15,000	
$10,000	$14,000

so you can counter the offer with "I believe my skills and experience are such that I should be in the $14,000-$18,000 range." By positioning your minimum near the top of their maximum, you encourage them to go beyond their original offer, either now or in the future. If the employer won't go to the figure you would like, request a three-to-six month review.

5. **Don't neglect fringe benefits.**
   These include insurance, pension plans, stock options, paid sick leave, child care, and so on. These are valuable, negotiable items.

6. **Get all agreements in writing.**
   Include promises for future raises. Offer to write the agreement yourself upon being offered the job.

NEGOTIATION TECHNIQUES

Negotiation techniques are limited only by your creativity, imagination and knowledge. At the same time, a job applicant's negotiation power is often limited by the set policy and procedure of each company (in terms of money, vacation time, sick leave, and fringe benefits they offer). However, there are techniques whereby you can obtain maximum settlement, feel good about the bargaining, and establish the potential for gain.

Below is a brief list of some techniques you can use when discussing salary, promotion, or just about any negotiable topic.

Multiple Choice Package. The more ways you give people to say "yes," the more chances you have of getting what you want. Give alternatives to consider rather than just a yes-or-no package. These may include higher upfront money, more fringe benefits, a higher performance bonus, or stock options.

Patience. Instead of an on-the-spot answer, suggest taking time to think it over or a "cooling off" period.

Back Burner. If you get a rejection, suggest putting the whole thing on hold for a few days. Remember, people adjust slowly to new ideas.

Plant Seeds. Keep planting hints, and suggestions, over and over again. Instead of giving up if you don't get everything, chip away at little things and at least get a small share.

A Slice At A Time. Just reach agreement on one thing at a time, and before you know it, you may have the whole bundle (or close to it).

Moon and Stars. Ask for more than you realistically can expect to get. This gives you room to negotiate. But remember: pigs get fat, hogs get slaughtered!

Deadlines. Establish limits and timelines, but be flexible if you see progress.

Hand Over Mouth. Silence is the most powerful tool you have. Just don't fall into the trap of filling theirs.

Surprise. Employ a sudden shift in your style. This may range from subtly varying the voice tone to dramatically changing from a calm manner to an explosive outrage.

Disposables. Include items that you expect to throw away so that the other side will see that you are giving.

Yes...and. Turn arguments to your advantage by agreeing with the other side, then overcoming the objection. Example: "I know I don't have a PhD, which the job description called for...let me show you how I used that time in the field accumulating just the experience that will help you solve your problems."

Underdog. Show your weakness, beg ignorance: "I need your help."

Call-in-the-experts. Use statisitics and the recommendations of authorities to support your claim.

Create Power. Get outside participants to go to work for you, have co-workers write letters of praise, compliment you to your boss, etc.

Bluff. Fold your papers, threaten to walk out, but be prepared to follow through.

Temporary Arrangements. Accept a less acceptable solution for a specified amount of time. Example: "I'll accept $1500 for 6 months, on the condition it will be raised to $1800 if you're satisfied with my work."

Negotiation is important not only to win a good salary, but your bargaining style will establish the way you expect to be treated. With preparation, practice, and the proper attitude, you should be able to bargain with confidence and finesse, paving the way to a satisfying and rewarding work experience.

SUGGESTED READING
Chapter 7

Chastain, Sherry, **Winning the Salary Game.** *New York, John Wiley and Sons, Inc., 1980.* Learning to assess the employer's needs and other valuable techniques for using negotiating know how.

Cohen, Herb, **You Can Negotiate Anything.** *New York, Bantam Books, 1982.* Fun to read and straight to the point. A best seller that talks about how to become a better negotiator.

Kennedy, Marily Moats, **Salary Strategies.** *New York, Rawson, Wade Publishers, Inc., 1982.* Tells the reader what they need to know about negotiating to stay in command. A book of tips and strategies.

Nierenberg, Gerard, **The Art of Negotiating.** *New York, Cornerstone, 1981.* This book explains the negotiation basics and effective strategies. Excellent for beginner or the more sophisticated bargainer.

CHAPTER 8

ON THE JOB

No matter how terrific you are, if no one else recognizes your contribution, you are in trouble.

Breathe a big sigh of relief and pat yourself on the back. You got the job! However, surviving, thriving, and achieving your goals is another story. To do so requires that you plot your course artfully and politically. Most job hunting books stop short of discussing how to begin a new job. Once you are hired, you are on your own. But we feel that just as important as finding the right job is keeping it.

More careers have been thwarted and ambitions dashed by believing the time-worn myth that "if you keep your nose to the grindstone and do your job well, you will be justly rewarded". Adele Scheele, in her book *Skills for Success*, discusses "Achievers vs Sustainers". Achievers are the people who spend 50 percent of their time working at the job and the other 50 percent creating opportunities for visibility and success. Sustainers, on the other hand, spend 70 percent of their time doing the job and the other 30 percent of their time wondering why they haven't been promoted.

To succeed, top quality work is absolutely essential for success. But in this highly competitive world of ours, competence and hard work in themselves no longer guarantee success.

Jodie's story illustrates what happens to a Sustainer. Jodie worked for a large industrial manufacturing firm as a marketing representative. She met her quotas, even earned a few bonuses, but watched as others who had been with the company as long as she were promoted while she stayed put. She couldn't understand it. She had worked long, hard hours, had taken seminars to upgrade her skills, but no one seemed to notice. However, ask her if she socializes with anyone outside her immediate coworkers, or if she has volunteered for any extra projects, and she will look at you in amazement. "Who has time?", she answers. As for documenting and communicating her ambitions and accomplishments, she assumes management knows.

Likewise, the reverse can be just as damaging. Joe is an

example of a brash overachiever. When Joe was hired from a competing firm to become the new regional sales manager, his top priority was to take the division out of the red in the first six months. To do so required sweeping changes. He immediately cut expense allowances, readjusted the incentive schedule, and demanded monthly progress reports. Joe knew his people would groan, but it was the only way to improve production. The eventual results were disasterous. Productivity plummeted along with staff morale. In an organization that valued team efforts, Joe made a poor impression.

Both cases portray people who failed to astutely assess their situations and proceed with savvy. The key to success is to move slowly but strategically. To help you get off to the right start, we have compiled a short list of suggestions. For more indepth advice, we urge you to look at the books listed at the back of this chapter.

PROCEDURE

Take at least six months to orient yourself. Immerse yourself in the company. Drawing attention to yourself too soon, like Joe did, could threaten insecure colleagues and supervisors, and jeapordize your relationships with them. Begin quietly, but not passively, like Jodie. Study the organizational chart becoming familiar with the distribution of power by job title. Familiarize yourself with key departments and people. Try to discern who has clout, remembering that power radiates through a web of informal relationships. Notice who lunches and socializes with whom. Get a sense of personalities, preferences, and the corporate climate. In other words, play detective and snoop around.

In addition, become acquainted with the purpose, plans and policies of the organization. What are the goals and objectives of the company, your department, and your boss? Everything you do must mesh with those goals.

Set goals and establish priorities. Determine how high you want to climb and in what direction, so you can make strategic choices and gain a sense of fulfillment as you progress. Make sure that your goals mesh with the organizational philosophy. To ensure that your ideas are compatible with the firm's values, study the company's long-range plans and seek help from those whose judgment you respect. Role models and mentors can be particularly useful as you establish your direction within the company, while sensitizing yourself to its values. A role model may be a person who has developed a career path in the company that you would like to follow. You might model yourself after the way they dress, communicate, or do business. A mentor, on the other hand, is an individual (usually an executive) who becomes your guide and your confidante.

Communicate your goals. You must ask for what you want, since your boss will probably not be a mind reader. It will be up to you to make your ambitions known. However, be sure that you are reaching the person who has the power to make decisions. Recognizing the individuals with the real clout in the organization is important from the start.

Cultivate relationships with everyone. If we were pressed to limit ourselves to one recommendation, this would have to be it. More than anything else, your relationships with your colleagues will determine your job success. Studies reveal that personality problems, not lack of skill, account for more than 77 percent of all job firings and resignations. Make a concerted effort to be on good terms with *everyone* in the company, from the janitor to the president, and especially the secretaries. Stay far, far away from feuds, conflicts, and cliques. And extend your social network out into the field: cultivate outside contacts by maintaining relationships with clients, joining professional organizations, and so on.

From the very first day, talk to and ask questions of every employee even remotely connected to you. Inquire about their jobs, preferences and expectations. Ask for their support and offer yours: "If you have any problems and I can help, I hope you will come to me. At the same time, I hope I can count on your help and support."

Your network is of prime importance. To lose peer support is to commit professional suicide. You leave yourself wide open to unpleasant surprises when you give up your access to the grapevines. Build trust and rapport by helping people out, doing favors, supplying extra little niceties, and asking for advice.

Practice these tools and techniques for reaching your goals.

1. **Pay careful attention to your appearance.** Your dress, materials, desk, and reports all should say, "I am a professional." Your image can literally make or break your chances. Be scrupulous in grooming it.

2. **Document your work.** From the beginning, maintain a portfolio that documents everything you do: ideas im-

plemented, compliments received, seminars attended, suggestions given, ways you helped your boss, and so on. Such a list will be useful when you need to support your request for a raise, a promotion, or additional benefits and responsibilities.

3. **Showcase your skills and talents.** .No matter how terrific you are, if no one else recognizes your contribution, you're in trouble. Visibility is essential. Volunteer to do extra work. Whatever you promise, deliver more. This can mean volunteering to start a task force to improve morale, showing up two minutes early to work, or completing a project ahead of time.

4. **Continually upgrade your skills.** To increase your worth, take courses, ask questions, attend conventions, and read trade and business publications to constantly update and expand your knowledge. Join professional organizations to gain additional opportunities for visibility and for demonstrating your talents. Work on developing leadership skills and gaining recognition and add these new accomplishments to your performance portfolio. You might even request that a letter of commendation for a job well done be sent to your boss by the president of the organization.

5. **Respect the chain of command.** If you consider going over the boss's head, risk only what you can afford to lose. A pecking order prevails. Consider using the informal system to reach the higher-ups by dropping the information you want to convey. All the while, however, be wary of being tagged too politically cunning. You may have to wait longer to achieve your objectives, but you will avoid becoming labeled as "too cunning."

6. **Delegate.** Finally, in your zealous effort to perform your job well, don't hoard all the responsibility. Don't let yourself get buried under the to-do piles. Overwork leads to burnout and robs you of valuable networking time. Moreover, making yourself indispensable, though it feeds your ego, could discourage your boss from losing you through promotion.

IN CONCLUSION

Jodie and Joe were able to salvage their jobs through career counseling. Each learned to apply the strategies outlined in this chapter to their situations.

Jodie decided to expand her network by developing contacts with co-workers outside her department and by identifying "rising stars" to serve as role models. She also began to use the work record included at the end of this chapter to document her job accomplishments. This not only helped her apply for future raises and promotions, but also boosted her self-confidence.

Joe, on the other hand, realized that he needed to loosen up his leadership style and listen more than dictate. He held weekly sales meetings to get ideas, complaints, and suggestions, and made sure his staff understood his reasons for the changes in policy. By taking time to nurture relationships, Joe was able to achieve success.

With careful planning, you too can achieve success on the job and lay the foundation for a winning future.

ON-THE-JOB ACTION PLAN

Just as you developed a written plan of action to research the job market, we suggest you follow a plan to establish goals for your career. However, keep in mind that a plan must be constantly evaluated and modified to adapt to changes that invariably occur.

Organization _____

Current position _____

What do I want to accomplish?

    In the next 6 months to 1 year:

    _____

    _____

    _____

    In the next 2 - 3 year?

    _____

    _____

    _____

How will I accomplish my goals? (tools, strategies employed)

    _____

    _____

    _____

Individuals who could support and influence me are:

    _____

    _____

    _____

Community resources I could use are:

_____

_____

_____

Obstacles I may encounter are:

_____

_____

_____

Strategies to overcome these obstacles are:

_____

_____

_____

A TOOL FOR RECORDING YOUR ON-THE-JOB WORK EXPERIENCE
(for each volunteer or paid position)

The completion of this kind of record can be your first step in putting together a work portfolio. Use a different record for each job title or position you hold within your company and/or professional and community organization.

Job Title/Position _____ **Dates Held** _____

Agency or Organization _____

Supervisor/Manager _____

Brief job description of duties and responsibilities:

_____

_____

_____

Skills and abilities used:

_____

_____

_____

Accomplishments/recognition:

_____

_____

_____

Likes and dislikes:

_____

_____

_____

What I learned about myself:

_____

_____

_____

SUGGESTED READING
Chapter 8

Kennedy, Marilyn, *Office Politics: Seizing Power, Wielding Clout.* *Chicago, Follet Publishing Co., 1980.* Who can bestow power? Who can take it away? How to use the office grapevine productively and make informed career decisions. A no-nonsense book.

MacKenzie, Alex and Waldo, Kay, *About Time! A Women's Guide to Time Management.* *New York, McGraw-Hill Books, 1981.* No-fail techniques for work and home tailored to the woman of the 80's but certainly just as applicable to men.

Scheele, Adele, *Skills For Success.* *New York, William Morrow and Co., Inc. 1979.* The author's special career strategies can help the reader realize their potential and reach their goals. An excellent book on job advancement.

Potter, Beverly, A., *Beating Job Burnout.* *San Francisco, California, Harbor Publishing Co., Inc., 1980.* Shares unique and useful techniques for preventing burnout on the job and renewing enthusiasm if it has already occured.

66 *I'm a great believer in luck, and
I find the harder I work the more
I have of it.*"

— Thomas Jefferson

CHAPTER 9

NEGOTIATING A RAISE

Preparation is the key to success and the best antidote for fear.

As an employee with (hopefully) a good record, you now have a lot more bargaining power than when you first negotiated your salary. Your employer will have sunk time, money, and energy into training you, so recruiting, hiring and training someone else would be quite costly. Moveover, your boss probably knows that a dissatisfied employee is not as productive as a satisfied one. Therefore, you need to assess your value to the company, inventory your skills, and evaluate your contributions. Then, pinpoint the right time to ask for a raise or promotion. Making a request for money isn't easy. On the contrary, it requires a lot of courage. Our standard advice still holds: preparation is the key to success...and the best antidote for fear.

The suggestions in Chapter VII for negotiating a salary are applicable to negotiating a raise. However, several additional recommendations are relevant here:

MATERIALS

Put together the following:

1. **A portfolio.** Include examples of your achievements, projects, and performance appraisals to document your case and illustrate your worth to the company. Also include letters of recommendation and thank-you's for jobs well done (by all means solicit these from co-workers and colleagues).

2. **A list of your duties.** Highlight anything you're doing in addition to what is expected of you. Include ways your responsibilities have expanded since your last pay increase. Also include how you helped to take over tasks from your superior and helped them look good.

3. **A market survey.** Find out what other firms are paying for the services you provide.

4. **Specific examples.** Any cost-effective activities in which you have helped the company earn or save time or money

(even if it was a more efficient use of paper clips!)
Everything counts.

PROCEDURE

Prepare and practice in the same way you would for salary negotiation.

Timing is critical. Pick a time when the company is doing well and/or your boss has had a success and, psychologically speaking, may be feeling generous. Never approach your boss on hectic days. Mondays and Fridays are not good days on which to ask for favors. Choose a time after you have had a recent accomplishment or have been given a new responsibility. Some experts suggest that you request a raise a few weeks before a performance appraisal is scheduled.

Assess your boss'_ needs as well as the company's. Is he or she seeking prestige, personal power, and/or promotion? If so, address these in your presentation.

Prepare a memo either for mailing prior to the meeting or to be left at the end of the talk, outlining what you want. This note will help your boss remember the issues and can be conveniently passed on to anyone else responsible for making final decisions.

Don't barge in on your boss unexpectedly. (Make an appointment)

Ask for a specific amount–don't merely request a raise. Back up the request with references to your accomplishments. Consider asking for a bonus if a raise seems out of the question at the time. You'll still be gaining something positive.

Project an efficient and competent image that says "I can help you; I'm worth more."

Listen carefully to objections. Keep your radar tuned in to unspoken ones. As long as they remain unexpressed you will be unable to overcome them. If you sense an objection, encourage your boss to voice it, and then counter it in a self-assured manner (this is where practicing with a "sparring partner" will have helped).

If successful, get the promise in writing. In fact, type up the agreement yourself.

If unsuccessful, find out why. Use the feedback in adjusting your objectives, activities, and perhaps your approach. Ask for

for another review in a specific number of months. That will be an incentive to set new goals and attain them.

WARNING

Never threaten to quit if you don't get a raise. That approach would only antagonize the boss and work against you. Instead, emphasize how much you enjoy your job and the company, and then raise the issue of adjusting your salary.

Don't answer, "I need it" when asked why you're asking for a raise. Needing more money isn't reason enough to get it. You must show that you deserve a raise based on your contributions to the company. Support your purpose in asking for a raise with facts!

Don't appologize, or appear unsure, or talk about how difficult it is for you to discuss money. These conversational tics make it easy for the boss to turn you down.

One of the major reasons people are underpaid is that they are simply afraid to ask for what they deserve. If you're nervous about requesting a raise, ask yourself what would be the worst thing that could happen if your boss said no. Once you have mustered the courage to approach your boss, prepared a good case, and understand the principles of negotiating, you will have built the self-confidence that will produce results.

66 *Behold the turtle . . . she only makes progress when she sticks her neck out.*"

— Anonymous

CHAPTER 10

WORK ALTERNATIVES

Consider these alternatives not only as viable options to traditional employment, but as a means of getting your foot in the door.

Up to now we have been discussing the typical 9 to 5 kind of job. However, you have other choices: temporary work, part time, flextime, job sharing, volunteerism, internship and starting your own business.

Consider these alternatives not only as viable options to traditional employment but as a means of getting your foot in the door, gaining experience, exploring new career fields and organizations, brushing up on skills and building credentials. They also provide survival income while you are looking for something more permanent. They are becoming incentive options for companies because of plunging morale and budget costs.

WHERE TO FIND

Of course, some businesses are more responsive than others. These would include companies that are:

1. Small and may not need full-time help.
2. Innovative and receptive to new ideas.
3. Low budget (educational, philanthropic, religious institutions).
4. Media firms that employ free-lancers.
5. Affected by peak periods (farms, accounting firms, mail order houses).

However, any employer may be a likely candidate. You may just need to work harder at selling the idea. Let's take a closer look at the options.

Temporary Work

When businesses find themselves needing extra help, but not wanting to hire permanent staff, they often turn to temporary employees. You can sign up with agencies found in the Yellow Pages under employment or newspaper want-ads specializing in this service. If you can type, you can always obtain a job. But aside from office work, there exists a tremendous variety of opportunities which range from promotions at trade shows, home health care, bartending, etc. Often these result in permanent positions. Temporary work provides an opportunity to get your foot in the door of a company and be there when job openings arise. Consider Elaine who wanted to make a transition from high school English teacher to a management trainee in industry. She found these trainee positions to be rare and hard to obtain, so to supplement her income, gain some familiarity with the corporate world and make valuable connections, she became a temporary receptionist for a variety of firms. She also became friendly with the directors of personnel and of training at the companies where she worked and let them know of her job objectives. After six months, one of the companies hired her.

Part Time, Flextime and Job Sharing

Over the past decade, many companies have begun experimenting with creative work alternatives. Three widely used options are:

Part Time Work. The latest census shows us that one out of every six persons is employed in jobs of thirty-five hours or less a week. In fact, the number of part time jobs is growing double the rate of full time positions. While the wages are usually low and fringe benefits lacking, this is changing in many areas, and part time jobs can be found in just about every profession.

Flextime. An alternative to the more rigid 9 to 5 day, the flexible work schedule allows greater influence over the hours worked. An employee agrees, for example, to be at the office during a core period of say, 10 a.m. to 2 p.m., and then can use his or her discretion as to which four hours to work.

Job Sharing. This involves an arrangement generally between two people who agree to share the responsibilities, salary and benefits of one job. In effect, one full time job becomes two part time jobs.

You may need to create a position if no such arrangment exists in the company you wish to work. Just as in the job interview, the key to winning approval for your proposal is showing the employer the advantages to him or her and the company.

Advantages could include: a positive alternative to lay-offs, a way to cover peak periods of business, to reduce overtime, to increase productivity and employee morale, to generate new options for older employees, to reduce absenteeism, and to retain valued employees.

To help you prepare for the possibility of seeking an alternative work option in your present job search or in the near future, we suggest the following steps toward creating an effective proposal:

1. **Study the new concepts of alternative work options to see which one satisfies your needs and meets your career goals.**
   - Read books and periodicals (see book list at the end of chapter).
   - Contact the National Job Sharing Network, listed in chapter resources, for information on work options in your locale.
   - Talk to others who have been involved in work alternatives either as employees or employers. (Use the informational interview to learn more about options.)
2. **Research particular industries and companies of interest to you in your career search.**
   - Make sure you understand the nature of the business.
   - Identify employer's personnel needs.
3. **Know yourself and your skills and abilities.**
   - Identify the personal and functional skills you have that will match the employer's needs.
4. **Develop a written plan to present to an employer.**
   - Describe your purpose in seeking a work alternative.
   - List the benefits of your proposed plan in meeting the needs of the employer.
   - Work out a proposed schedule – that is, a list of tasks and responsibilities plus a wage and benefits package.
5. **Create a marketing strategy for contacting employers.**
   - Use the techniques described earlier in this book for researching and networking.
   - Practice your techniques and work with a friend on your presentation.

Be aware of the fact that many companies are unwilling to take a risk and implement a new staffing plan. All the more reason to thoroughly research, prepare and sell the benefits of yourself and your plan to the company. Making the company's administration knowledgeable about creative work alternatives may be step one.

Many of our clients have been successful at creating alternative work arrangements. A good example is Tom, a job seeker who was making a career change from the government sector into business. His job objective was to find a position that allowed him to work in organizational long-range planning. After a long period of networking, investigating and interviewing, John was referred to the president of a commercial architectural firm who was looking for someone to do public relations work, which included writing brochures and trade journal articles. Tom had not seriously considered a part time job as anything but temporary, nor had he thought of public relations as a possible way of emphasis. However, after many conversations with executives of the firm and others in the marketing field, Tom began to realize that in this position, he would be developing new areas of expertise in marketing that could eventually lead him to long-range planning. In addition, he would be creating an identity for himself in the business community that would be important when he made the change from the public to the private sector.

The moral of Tom's story to the unemployed job hunter: consider part time employment not only as a stepping stone to a full-time job but as a productive means of keeping your skills fresh and active. After only two months, Tom proved his potential and now has a full-time position at the architectural firm as marketing director, a job he created.

Another example is Sue and Joanie who were employed as administrative assistants on a job sharing basis. Each had different areas of expertise and interest – Sue in dealing with clients and projects, and Joanie in organizing and maintaining a research library. The job sharing arrangement satisfied the boss's need to have his office staffed daily. In addition, it allowed each woman to capitalize on her individual areas of skill and interest. Another advantage was that during peak times, vacations and sickness, one worker could easily replace the other.

Volunteer Work and Internship

Non-profit community service organizations are starving for help. But do not eliminate profit-oriented companies. There it is called internship, interning being akin to apprenticeship. If you are considering changing career fields or re-entry into the job market, you don't have to be highly skilled in a particular area such as public relations to volunteer your time and in return build skills, credentials, experience and contacts. Volunteer work is often the ticket to paid employment, and should be treated with the same professionalism. Keep records, maintain a portfolio, obtain letters of recommendation and by all means include it on your resume under employment history.

Susan was working on her MA in Guidance and Counseling and had not decided on a particular emphasis in her studies, when she decided to attend a seminar offered by the University Women's Center. In discussing her career plans with a small group of other women, she began to realize that she needed to have some practical experience in different areas of counseling before she could decide on a specialty. She signed up as a volunteer at the Women's Center learning to design and facilitate career seminars for women. It was this training experience that convinced her to choose a counseling emphasis in career development. Not only did she have an opportunity to further her own training, but the volunteer experience provided her with visibility, contacts, and self-confidence.

Entrepreneurship

Starting a business seems to be the American dream. And, in these days of sparse job openings, this option is becoming more and more popular.

However, we feel obligated to caution you first. Building your own business is no quick cure for the job-hunt jitters. The time and energy necessary to research and build your business will make job hunting look like a vacation. The approach is basically the same: advance preparation and planning, research and networking. Moreover, in the first one to three years, you will probably work longer hours (expect 12 hour days) and earn less than your salaried friends.

Financial assistance is frustratingly difficult to obtain (some say impossible without a rich relative or a generous friend). To further dampen your spirits, the Small Business Administration claims that thirty-three percent of new businesses fail in their first year, fifty percent fail in the second, and only thirty-four percent make it past five years.

However, the rewards of being your own boss can outshine the negatives if you do your homework (do we sound like a broken record?).

Below we have briefly sketched out the basic steps in planning self employment.

1. **Learn all you can** about yourself, your field and business ownership in general. That means:
   - *Discuss your ideas with friends, colleagues and authorities in your field.*
   - *Talk to business owners.*
   - *Study the competition (include other geographical locations).*
   - *Attend workshops, seminars, conventions.*
   - *Read books, journals, newspapers.*
   - *Join a professional network of other business owners to give you motivation and support.*
2. **Find a need and fill it** (or create a need if you have to). Pick a field you know and love.

3. Examine all options. For example, will you:
- *Start your own shop or office or work out of your home?*
- *Acquire a going firm or a franchise or start from scratch?*
- *Moonlight until you can go full time or plunge right in?*
- *Share a partnership or become sole owner?*

4. Develop a comprehensive written plan:
- *Describe your service or product.*
- *State your mission, goals, and objectives.*
- *Pinpoint your expected sources of business and income.*
- *Identify your competition (present and potential).*
- *Describe the advantages of using your business over the competition.*
- *Develop strategies: marketing, pricing, expansions.*

5. Stop here ... and make sure all these make sense and are realistic and attainable.

6. Get Experience.
- *Take a temporary job in the field.*
- *Volunteer in the field.*
- *Take relevant college courses (since ninety percent of the failures are due to managerial incompetence, be sure to take a management course).*
- *Consider related training programs.*

7. Employ professionals.
- *Having an accountant and a lawyer is a must.*
- *A banker and insurance agent are important.*
- *Using consultants with special expertise in your business area may be a wise investment.*
- *Know your limits, and delegate. Hire help if you need it.*

8. BELIEVE WITH ALL YOUR HEART YOU WILL SUCCEED!

WARNING: You will be confronted with many roadblocks, including people who tell you your idea will never fly. This is important for you to hear. Confronting these pessimists will force you to test your convictions and desires and the level of your commitment. Realize that they are only people with opinions, not prophets with psychic powers. If you have done your homework, and passionately believe you can succeed, then by all means DO IT, and good luck!

SUGGESTED READING
Chapter 10

Alter, JoAnne, **Part-time Career for a Full-Time You.** *Boston, Houghton Mifflin Co., 1982.* Nuts and bolts guide for anyone who seeks a good part time or shared job. Especially good for those with little job hunting experience.

Mancuso, Joseph R., **How to Start, Finance and Manage Your Own Small Business.** *New Jersey, Prentice-Hall, 1978. Step-by-step guidelines and business plans to use. Excellent coverage of the topic.*

Meier, Gretl S., **Job Sharing: A New Pattern for Quality of Work and Life.** *W. E. UpJohn Institute for Employment Research, 300 S. Westnedge Ave., Kalamazoo, Michigan, 49007, 1978.* Discusses how to present a job sharing proposal and gives case studies on its use.

First National Directory on Part time and Flextime Programs, *National Council for Alternative Work Patterns, 192S K Street N.W., Suite 308A, Washington, D.C., 20036, 1978.*

A Guide to Splitting or Sharing Your Job. *New Ways to Work, San Francisco, California, 1981.* How to negotiate with an employer, how to find partners, sample proposals.

New Ways to Work. *149 Ninth Street, San Francisco, California, 94103.* The oldest and most experienced work time programs in U.S., a not-for-profit community-based work resource organization that was established in 1978 to respond to workers' and institution's needs; good resource for selected bibliographies on the subjects.

66 *If we do not change our direction,
we are likely to end up where we
are headed.*"

— Chinese proverb

CHAPTER 11

HELPING ANOTHER

Supporting without intruding, nudging without pushing, lending your ear but closing your mouth takes a great strength.

What can you do to be truly helpful when someone you care about – a friend, spouse, or child, for instance – is looking for a job? To facilitate the solution and avoid being part of the problem, fight off the temptation to do their work for them. While you may sincerely want to do as much as possible to ease the other's plight, you could end up doing harm by doing too much.

Still, if you read this book and understand the job-hunting process, without pushing, tugging, or controlling, you can help another through this difficult period while allowing them to develop the skills necessary to progress in their career. It's not easy to watch another struggle. Supporting without intruding, nudging without pushing, lending your ear but closing your mouth takes great strength. But, by acting as a sounding board, the results will be better in terms of their success and your relationship.

You want to:

1. Offer support and encouragement.

2. Help the job seeker talk through pressures and obstacles.

3. Encourage them to discuss ideas.

4. Provide requested (but not unsolicited) information and referrals.

5. Discuss the self-awareness exercises and job-hunting procedures.

6. Foster confidence in their own judgment and inner resourses.

7. Role-play possible interview situations.

You know that job hunting can be frightening, that self-assessment can be distressing, and that change itself is discomforting. Your prime task is to reduce the stress and confusion that may be fogging the job seeker's thinking. No one can make sound decisions under emotional duress.

To do this, let the person talk while you listen. Like removing a lid from a pressure cooker, talking permits the release of internal pressures and pent-up emotions. By listening with understanding, you provide a safe space in which troubling issues can surface. Because confusion has a way of blinding us, your sensitive responses will not only give solace and support, but also will serve as a mirror reflecting back useful thoughts and sentiments.

Paraphrase and summarize the other's ramblings, enabling him or her to hear what was said. Such phrases as, "I hear you saying..." and "It sounds like..." will help a confused talker clarify muddled thinking. Furthermore, by listening and reflecting the talker's ideas, you will be creating a bond that builds trust and support and paves the way for your working together.

Resist the temptation to bombard job seekers with ideas of what career they should pursue. No matter how constructive your suggestions or sincere your intentions, at the initial stage, the person in the throes of change will lack the emotional perspective and necessary data to make wise choices. Likewise, refrain from lecturing. Whatever your knowledge and experience, and however close your relationship, it is unlikely that you will really know what's best for another person. Each of us must reach our own conclusions. Your function is to gently guide, encourage, and befriend.

PROCEDURE

When you sense their confusion lifting and desperation easing, begin helping them to gather information on which they will base their decisions and direct their efforts, but don't push them. Don't rush them. We suggest working with them through the step-by-step procedure presented in this book:

1. **Begin by explaining the process of career planning.**

2. **Explain the importance of self-awareness.** Encourage them to do the exercises. Then make yourself available to discuss the results with them.

3. **Help them develop a job objective.** Review what they have learned.

4. **Encourage informal interviews.** Assist them in drawing up a list of potential contacts.

5. **Encourage informational interviews.** Role-play for practice.

6. **Discuss their progress.** Praise any small efforts. Ignore setbacks. Don't nag or make a big deal because

they haven't had an informational interview for three days.

TWO ADDITIONAL POINTS ARE RELEVANT:

7. **Don't take their emotional outbursts personally.** Under a great deal of pressure, they may lash out at the most convenient target-probably you. Protect yourself while allowing for their release. To relieve your stress, get involved in support groups, physical exercise, journal writing, or talking with friends or a counselor.

8. **Don't push.** Transitions seem to have their own pacing mechanisms. Often, people in pain retreat to lick their wounds before they can go out into the world once more. Nagging will only be like rubbing salt in open sores.

If you suspect that the person is "hopelessly stuck," here are some suggestions:

ENCOURAGE CHANGE IN OTHER AREAS. Forget the job and suggest that they do something different: make a change in their appearance, take a course, drive along a new route home - anything new and different. Change has a ripple effect and can be profoundly energizing in all areas of life.

GIVE SUBTLE, GENTLE NUDGES (NOT SHOVES). Arrange for another, more objective friend to help. Suggest a job-hunting workshop or support group or an article or book on the subject.

RECOMMEND COUNSELING. Local colleges or mental health centers provide assistance for small (or no) fees.

You can see there is much you can do to help another, short of actually locating a position for them. By being knowledgeable about the techniques of job hunting, trusting the other's ability to make the right decisions, and supporting his or her efforts, you can bolster their sagging confidence and serve as a guide through this difficult transition to becoming happily employed.

66 *A thorough, professional job search can be the best investment you'll ever make.*"

— A former job hunter

CHAPTER 12

TAX TIPS FOR THE JOB HUNTER

Only those expenses directly related to job hunting may be deducted.

The goal of this chapter is to make you more aware of the limits and opportunities available to you when filling out your tax return while job hunting. This information is not intended as legal or financial advice, but is derived form government documents and court decisions which have clarified the tax code. We assume no liability for actions taken as a result of information contained herein. See your lawyer, accountant, tax advisor, or IRS publications for further help.

WHO QUALIFIES

You may declare job hunting expenses as a tax deduction only if you are seeking employment in the same trade or business and you itemize deductions on Form 1040 (long form). You may declare these deductions even if you do not change jobs.

Don't be too put off by that last requirement. Court rulings have defined the same trade or business as services performed...that is, your job classification. For instance, if you are an administrator for a computer manufacturer, and you want to find a job as an administrator for a newspaper publisher, you qualify. It's not your employer's business that determines your occupation but the nature of the work you do.

So, if you are an engineer working on bridge construction, you are not changing your "trade or business" if you seek employment as an engineer in the shipping industry. The IRS should allow all job hunting deductions under these circumstances.

But, if you (the engineer) decided to look for work as a gourmet chef, the Internal Revenue Service would presumably deny the deduction for job hunting expenses. They would also deny the deduction if you have not established a reasonable continuity in your line of work.

WHERE TO DEDUCT

The other hitch is that you must be able to itemize deductions on Form 1040. Some people do not have enough medical expenses, tax payments, interest expenses, contributions, casualty or theft losses, or other miscellaneous deductions to use Form 1040. They must take a "standard deduction" on the long form. Check your tax return instructions to see if you can itemize.

If it appears you are able to take itemized deductions, then your job hunting expenses should be recorded under Miscellaneous Deductions at "Expenses of looking for a new job". Most miscellaneous itemized deductions can be deducted only to the extent that they exceed 2% of your adjusted gross income.

Some of the deductions subject to the 2% limitation are as follows:

- Expense of looking for a new job
- Cost of work-related small tools and supplies
- Qualified educational expenses
- Certain work clothes and uniforms
- Tax counsel and assistance
- Professional society dues
- Office-in-the-home expenses

(Many of the above noted expenses have certain restrictions and qualifications in order to be deductible.)

To the extent your job hunting expenses and other miscellaneous itemized deductions subject to the limitation exceed 2% of your adjusted gross income, they become "qualified deductions". You can determine if you use your itemized deductions or the standard deduction.

If you meet the above criteria, you can deduct your job hunting expenses even if you do not find a job or if you choose not to change jobs after looking for new employment. Even if you accept a job offer, then turn it down because your employer offered you a raise, you can deduct job hunting expenses. That's because your employer's offer is a direct result of your job hunting activity.

Is that all too confusing? Well, we tried to warn you. The IRS has so many if's, and's and but's, and or-else's that we got caught up in it all. Nevertheless, the information is accurate as far as we know right now.

WHAT TO DEDUCT

Let's leave **who** can deduct behind for the moment and turn to **what** you can deduct.

All travel and transportation expenses related to your job search are deductible if the nature of your travel was solely to seek new employment. You may deduct:

1. Commercial transportation fares
2. Operation and maintenance of your automobile

3. Taxi fares
4. Meals and lodging
5. Cleaning and laundry expenses
6. Telephone and telegraph expenses
7. Tips
8. Other travel expenses related to job hunting

If you combine a job-hunting trip with a personal trip, travel expenses to and from the destination are deductible only if the trip is related primarily to seeking new employment. The amount of time you spend on personal activity compared to the amount of time you spend seeking employment is important in determining whether the trip can be deducted or not.

Even then, only those expenses directly related to job hunting may be deducted. You cannot, under any circumstances, deduct expenses for activities which are considered entertainment, amusement, or recreation unless you can substantiate that they were directly connected to job hunting.

If you have your resume prepared by an outside service, these expenses are deductible. You may also deduct the postage used to send resumes to prospective employers.

Some people will find employment through an agency which takes its fee from the job hunter rather than from the employer. If your new employer reimburses you for this fee immediately, then it is not considered income. Nor is it considered income if your employer pays the fee directly to the agency.

But if the employer reimburses you only after a period of satisfactory service, then it becomes additional wages, and you must declare it as income. Of course, you can also deduct the fee you paid the employment agency. Whether your employer reimburses you or not, you may deduct employment agency fees, job counseling fees, and other related professional fees.

A prospective employer may reimburse you for expenses incurred while traveling to and from an interview. These are not considered income unless they exceed your actual expenses. But, the excess must be declared as income. If your interview expenses are picked up by an employer, you cannot deduct them, of course.

PROVE IT

If you are ever audited, the IRS will ask to see proof of your expenses including receipts, cancelled checks, and other evidence. We urge you to keep as detailed an accounting of your activities as you possibly can.

You must keep track of...
- the amount of each separate expenditure
- dates
- location of the activity
- purpose
- mileage for each trip

You should keep these records in an account book, diary, statement of expense, or similar record supported by adequate evidence which establishes the nature of the expenses. Although not mandatory, written, contemporaneous records are the best evidence to establish your deductions. So start keeping track of your expenses as soon as you begin job hunting.

Sufficient documentation, as far as the IRS is concerned, consists of a bill or receipt which shows the amount, date, place, and essential character of the expense. A lodging receipt with the hotel's name, address, dates of your stay, and itemization of meals, rent, and telephone is considered adequate.

A cancelled check or credit card receipt along with a descriptive bill will usually suffice. But, a cancelled check is not enough in itself to show that the expense was related to job hunting. The IRS will request some other form of proof.

You should not count on it, but the IRS may be willing to accept your personal word in lieu of a bill if you can describe details and provide other supporting evidence (newspaper ads, for instance). It's a long-shot and should only be used if you failed to keep track of receipts for a small part of your deductions.

Consult your tax advisor or the IRS for further information.

ONE FINAL WORD

We have attempted to show you a methodical approach to not only becoming happily employed, but to enjoying the journey along the way. However, we find there are some points we have not mentioned. Yet these are thoughts that we strongly believe can make a big difference in your efforts.

First recognize that you have a powerful ability to shape your future. There is a technique called autogenics. This simply means repeatedly and vividly seeing yourself in the job you want. Repeat to yourself and write down positive statements affirming the realization of your goals. Visualization and affirmations are profoundly potent tools.

Moreover, not only are you powerful; you are wise. Trust your hunches. If you are armed with good solid information, your best guide will always be your own inner signals. In your heart, you know better than anyone what you want and what you need to do. Listen to that inner voice. Trust it.

With all the instruction, advice, and guidance we have offered, it all boils down to YOU. You must do the work to make your success happen.

By taking the time to figure out what you want, to talk to other people, to research possibilities, and to use all resources available to you, you are well on the road to success. It is a rocky road, to be sure, full of potholes, bumps and dead-ends. You are bound to run into patches of fog and moments of utter darkness. It requires you to keep on going through mirky confusion and ambiguity, to be willing to risk failure and endure rejection without personalizing it and to keep up faith when there is no end in sight.

Nevertheless, it is a path well worth taking. Ask anyone who is happily employed. They will tell you about the joys of using your talents to their fullest, doing something that really matters.

Throughout these pages, we have tried to be your guide, give you some helpful pointers, and most of all, keep you going through the mire. To that effect, we would like to leave you with a short parable as a way to summarize the main theme of this book.

> *"When nothing seems to help, go and look at the stonecutter hammering away at his rock perhaps a hundred times without as much as a crack showing in it. Yet at the hundredth and first blow it will split in two, and you know it was not that blow that did it, but all that had gone before."*

WE WISH YOU SUCCESS ...

Barbara   *Janice*

ST. LOUIS AREA JOB HUNTER'S RESOURCE GUIDE

Researched, written and compiled
by Marmie Tuerff Edwards

Marmie Tuerff Edwards

Trained and experienced as a journalist, Ms. Edwards developed her knowledge of several major industries as a legislative aide on Capitol Hill. She researched and wrote this guide while in St. Louis after working as a publications editor at Washington University in St. Louis. Now back in Washington, she writes, edits and serves as a public relations representative at George Washington University.

HOW TO BECOME HAPPILY EMPLOYED IN ST. LOUIS

Few people look forward to a job search but with the abundant resources available in St. Louis, it can be an adventure rather than a drudgery. By using this guide, you need not experience the sense of "going it alone" that many job hunters face. You will find there is always another person to contact, another source to investigate, another lead to follow.

For those of you unfamiliar with the St. Louis area, there are three major geographical divisions: St. Louis (city), St. Louis County (in the west with Clayton the county seat and main business district), and East St. Louis on the Illinois side of the Mississippi River. As the area expands into the suburbs, Chesterfield, on the southwestern edge of St. Louis County, may eventually threaten to draw business away from Clayton in much the same way Clayton threatened the downtown area a decade ago. In the far northwest, St. Peters, said by some to be the fastest growing bedroom community in America, has drawn new industries attracted by the suburb's youthful, well-educated residents.

Given the size of this region and the limited space of this book, we can offer only a sample of what is available. We regret having to exclude any resources, but subjective and arbitrary decisions were necessary to keep the book to a reasonable size. In no way does inclusion constitute an endorsement, nor does exclusion indicate a lack of regard for any organization or service.

We intend the following information to offer you a starting point. After reading this book, we hope you are convinced that more than 75 percent of today's jobs are found through informal networking and personal effort. That is why we compiled this guide – to introduce you to some of the resources needed to make your job search easier and more productive.

Included here are listings of major employers, professional organizations, job search centers, and a sampling of employment agencies and career counselors. Throughout the book there are several items about little known services to aid your search. For example, did you know the downtown public library is one of the region's best sources for career information? Or, did you know that the community colleges sponsor support groups for women re-entering the job market? Are you trying to make contacts with a professional organization? Check through back issues of the *St. Louis Post Dispatch* Monday Business section for meeting dates and times.

Use the guide to stimulate your imagination. Call the sources in your specialty to generate new leads, new ideas. We would appreciate it if you'd share your ideas with us. If you find valuable resources that you feel should be in the next edition of the guide, please complete the form at the end of the book. We would enjoy hearing from you.

ST. LOUIS AREA JOB HUNTER'S RESOURCE GUIDE

Contents

CAREER INFORMATION AND JOB REFERRAL RESOURCES

Non-Profit Organizations

Whether you have ample financial resources or are running low on funds, many organizations will help you assess your talents, then teach you job-finding and resume writing skills. Job hunting can be a lonely process unless you can hook into a job hunters' network like one of those listed below.

Businessmen Between Jobs
601 East Claymont Dr.
Ballwin, MO 63011
(314) 394-2233

Middle managers meet each Monday morning at 9 a.m. to improve job-hunting skills and maintain organizational, administrative and leadership abilities. Mini-resumes are mailed to approximately 2,500 potential employers every six weeks. Fee: $5 lifetime membership and $5 charge to be included in mini-resume mailing.

Careers for Homemakers
St. Louis Community College
Florissant Valley
3400 Pershall Rd.
Ferguson, MO 63135
(314) 595-4565

Assists displaced homemakers, unemployed or underemployed become emotionally and economically self-sufficient through satisfactory employment. Program includes a workshop, group counseling and job-hunting techniques, weekly job club at Forest Park Community College. Meramec campus also has a branch.

Catholic Charities Job Bank
4141 Forest Park Blvd.
St. Louis, MO 63108
(314) 531-2259

A job-matching program open to all members of the community designed to expand job seekers' networks of job leads. Publishes weekly help and situation wanted ads in the St. Louis Review. Phone lines open M-F 9 a.m.-1 p.m.

Center for Career Development
2555 S. Hanley Rd.
Brentwood, MO 63144
(314) 647-5533

United Way agency providing short-term vocational counseling, testing as needed and a resource library equipped with the Missouri View computerized information system listing 598 career fields, plus applicable areas of study at state colleges. Special placement assistance for the handicapped. Fees: approximately $25 per session.

International Institute
3800 Park Ave.
St. Louis, MO 63110
(314) 773-9090

Multi-service agency assists refugees from politically repressive countries to resettle and gain self-sufficiency, includes identifying fields with employment potential and provides a job placement service. English as a Second Language courses available 9 a.m.-12 noon and 7-9 p.m. No fees to refugees. Non-refugees pay $50 for language courses.

Metropolitan Reemployment Project
5600 Oakland Ave.
St. Louis, MO 63110
(314) 644-9142

Provides resume preparation and counseling services to laid-off workers, teaches interviewing skills and vocational exploration, lists current job openings within Metro area. Vocational ed courses available in cooperation with local schools. No charge to those presenting letter from former employer.

Redevelopment Opportunities for Women
2709 Locust
St. Louis, MO 63103
(314) 531-6373
Provides a seven-week course in essential life skills to disadvantaged women 18 and older, offers group counseling and preparation for employment.

YWCA
140 N. Brentwood Blvd.
Clayton, MO 63105
(314) 725-7203
Developing a Women's Resource Center to offer career counseling and personal inventory testing with a certified counselor. Fee: $15 for members, $20 non-members for first visit.

YWCA of Metro St. Louis
1015 Locust St.
Suite 734
St. Louis, MO 63101
(314) 421-2750
Offers career counseling, networking and workshops for women.

Community Colleges and Universities

Several St. Louis area educational institutions offer continuing education courses in career planning. Others, particularly the community colleges, provide a wider range of career counseling, job search workshops, vocational testing, job referrals and resource libraries. If you are an alum of a local college, call their career planning and placement office to see what services are available. If not, check with the University of Missouri-St. Louis to see if they will establish a reciprocal agreement with your alma mater. If you are considering going back to school or branching into a new field, contact several colleges, private as well as public, to see which one will offer you the courses you want with the best financial aid package.

MISSOURI

Fontbonne College
6800 Wydown Blvd.
St. Louis, MO 63105
(314) 862-3456
Small liberal arts college known for home economics, speech therapy and individualized arts program.

Harris-Stowe State College
3026 Laclede Ave.
St. Louis, MO 63103
(314) 533-3366
Four-year college founded in 1857, offers one of the only B.S. degrees in urban education in the nation.

Lindenwood College
Kingshighway & First Capitol Drive
St. Charles, MO 63301
(314) 946-6912
Private liberal arts college offering business and graduate degree course work in the evening and on weekends, plus regular daytime degree programs.

Maryville College
13550 Conway Road
St. Louis, MO 63141
(314) 576-9300
Career-oriented liberal arts college with an active career planning and placement office.

St. Louis Community College System

Branches:

Clarkson Education Center
P.O. Box 1500
Ballwin, MO 63022
(314) 394-8600

Florissant Valley
3400 Pershall Rd.
Ferguson, MO 63135
(314) 595-4200

Forest Park
5600 Oakland Ave.
St. Louis, MO 63110
(314) 644-9100

Meramec Campus
11333 Big Bend Blvd.
St. Louis, MO 63122
(314) 966-7500

South County Education Center
3701 Will Ave.
St. Louis, MO 63125
(314) 894-0007

All five locations have career planning centers and offer continuing education courses for job enhancement, career development or career changes.

St. Louis University
221 N. Grand
St. Louis, MO 63103
(314) 658-2222

Large Catholic institution with career planning and placement services available only to students or graduates.

St. Mary's College
200 N. Main St.
O'Fallon, MO 63366
(314) 441-7040

Small private college expanding its medical and business offerings, some career planning services available.

University of Missouri-St. Louis
8001 Natural Bridge Rd.
St. Louis, MO 63121
(314) 553-5000

Large public university with career planning and placement for its students and alums, but also open to graduates of other universities on a reciprocal basis.

Webster University
470 E. Lockwood
St. Louis, MO 63119-3194
(314) 968-6900

Private institution known for its liberal and performing arts programs. Career planning and placement is limited to its own students.

Washington University
One Brookings Drive
St. Louis, MO 63130
(314) 889-5000

Private research institution offering a broad range of undergraduate and graduate degrees. Placement and career planning is limited to students and alums.

ILLINOIS

Belleville Area College
2500 Carlye Rd.
Belleville, IL 62221
(618) 235-2700

Continuing education courses in career planning and career exploration are available, call for details.

Southern Illinois University at Edwardsville
Edwardsville, IL 62026
(618) 692-2000

Contact the college for information about career planning and placement, as well as a list of available workshops.

State Community College of East St. Louis
601 James R. Thompson Blvd.
East St. Louis, MO 62201
(618) 274-6666

Career planning and placement services are available, call for further information.

Job Search Centers

Offices listed below offer job search workshops as well as resources for occupational information, unemployment insurance and job placement.

St. Louis Work Incentive Program
505 Washington Ave.
St. Louis, MO 63101
(314) 231-7348
Job-matching program and unemployment security program.

St. Louis Co. Dept. of Human Resources
555 S. Brentwood Blvd.
Clayton, MO 63105
(314) 889-3453
Assists disadvantaged with employment training and counseling, provides vocational and on-the-job training, and job referral. Applicants should call 423-8655.

St. Louis City Agency on Training & Employment
317 N. 11th St.
St. Louis, MO 63101
(314) 241-4300
Vocational training in welding, carpentry and liscensed practical nursing for city residents 18 and older.

Missouri Division of Employment Security
421 Dunklin
P.O. Box 59
Jefferson City, MO 65104
(314) 751-3215
State office if you need more information than the local offices can provide.

ILLINOIS

Illinois Dept. of Employment Security
646 N. 20th St.
East St. Louis, IL 62205
(618) 271-7750
Job placement, workshops and unemployment security

Illinois Dept. of Employment Security
4519 W. Main St.
Belleville, IL 62223
(618) 233-4735
Regional office offering full job/employment search capabilities, administers benefits to unemployed.

Job Training Partnership Office
512 East Main St.
Belleville, IL 62220
(618) 277-6790

Government Personnel and Services

More than 35,000 federal workers are employed in the Metropolitan area. Some work at the Arch for the National Park Service, some with the FBI, others with the Department of Defense or the Air Force at Scott Air Force Base. Listed below are several numbers where you can get more information about this segment of the St. Louis area's employment picture.

Federal Personnel Management
815 Olive St.
St. Louis, MO 63101
(314) 425-4285
Federal Job Information Center.

St. Louis Dept. of Personnel
City Hall, 1200 Market St., Rm 100
St. Louis, MO 63103
(314) 622-4308
Contact this office for information about jobs in the city of St. Louis, not in the county.

Equal Employment Opportunity Commission St. Louis Civil Rights Enforcement Agency
10 N. Tucker Blvd., Rm 300
St. Louis, MO 63101
(314) 622-3301
Enforces federal laws concerning discrimination in employment and housing and monitors city contracts to insure equitable hiring of minorities.

CAREER EXPLORATION AND TRAINING OPPORTUNITIES

Internships and Cooperative Jobs

Whether paid or unpaid, internships offer a structured work experience with supervision and evaluation, and usually require a specific time commitment. Cooperative programs, like internships, provide on-the-job training and evaluation, plus either a stipend or an hourly wage. A decade ago cooperative programs were federally funded with students alternating semesters in class or on the job. Today more co-op programs combine studies with work experience, so learning can be applied immediately in the workplace. Many opportunities exist for students at community colleges and universities. These are often listed with the campus career planning and placement center. If you are not enrolled in school, but would like to see how your talents match a specific field, it may take more determination on your part. But it can be done. Frequently internships in radio, television and with some ad agencies continue to be unpaid, but they offer the experience needed before a paying job can be obtained. In banking and architecture there are opportunities for co-op arrangements in St. Louis. Larger companies such as McDonnell Douglas, Monsanto, Olin and Emerson Electric have initiated formal co-operative programs with students at various universities.

Volunteer Organizations

The United Way operates the major volunteer bureau in the St. Louis area, but there are several other opportunities with cultural and health care agencies. Just a few are listed below. You can contact your nearby hospital if you have a few spare hours a month and want to see if that environment appeals to you. Or the Art Museum and the Historical Society can also use your help. Several of the organizations listed under *Survival Resources* would also welcome volunteers. Lend a hand, you'll meet people with similar interests and gain valuable experience as well.

Adult Education Council of Greater St. Louis
7711 Bonhomme, 4th Floor
Clayton, MO 63105
(314) 726-1210
A good source if you want to know if an unusual course is offered anywhere in St. Louis.

American Cancer Society
1280 Research Blvd.
St. Louis, MO 63132
(314) 567-9730

American Heart Association
4643 Lindell Blvd.
St. Louis, MO 63108
(314) 367-3383

American Red Cross
4050 Lindell Blvd.
St. Louis, MO 63108
(314) 658-2000

Arts & Education Council
40 N. Kingshighway
St. Louis, MO 63108
(314) 367-6330
More than 138 cultural and educational organizations in the Metro area are represented by the Council. Publishes a monthly calendar of events, a media guide, an artists registry and raises funds for the St. Louis Arts Festival.

Planned Parenthood Association
2202 S. Hanley Rd.
St. Louis, MO 63144
(314) 781-8300

Ronald McDonald House
4381 W. Pine
St. Louis, MO 63108
(314) 531-6601

Salvation Army, Family Services
3744 Lindell
St. Louis, MO 63108
(314) 534-1250

United Way/Voluntary Action Center
915 Olive
St. Louis, MO 63101
(314) 421-0700

Apprenticeships

If you would like to learn a particular trade or craft, you can attend classes at one of several technical training centers in the Metropolitan area. In order to become an apprentice (for example in the automotive, industrial or building trades), you must be accepted by an employer. However, by taking required course work at a technical center you can learn valuable skills in computer repair, carpentry, drafting, welding, machinery repair or industrial manufacturing. These skills can help you sell yourself to an employer and give you up to two-years credit toward a four-year apprenticeship program. Financial assistance is available for students enrolling in these programs. And once you are accepted as an apprentice, most employers will assume the cost of your schooling. Apprentices generally receive 60 percent of a journeyman's salary, moving up 5 percent every six months. You can contact your union for more information or check the Yellow Pages for technical schools. Do ask to talk with others who have completed the program before enrolling into one of these courses. Some programs are more successful than others in helping their students find employment. Call to check about the cost of tuition and fees that may change before publication.

Lewis and Clark Community College
5800 Godfrey
Godfrey, IL 62035
(314) 741-2779

Provides machinist and electrician apprentices with a supplemental educational program ranging from two to four years. Costs range from $325 per year for evening classes to $1,200-1,400 per year for full-time students.

Rankin Technical Center
4431 Finney
St. Louis, MO 63113
(314) 371-0236

Trade school offering a variety of specialities including computer repair, industrial, automotive, carpentry, drafting, welding and machine shop. Fees vary from $1,800 for a two-year evening program to $5,500 for a two-year day school program.

North County Technical Center
1700 Derhake
Florissant, MO 63033
(314) 839-6101

Offers a wide variety of trade programs both to high school students and through an adult education program in the evening.

PROFESSIONAL HELP
Career Consultants

If you decide you want some assistance in your career assessment and job search, and are willing to pay for it, determine what type of help you need before you invest your money. Some career firms offer only vocational testing and resume preparation, while others guide you through a complete self-assessment and the entire job search. Fees vary greatly and are usually paid by you, not your employer.

Choose a consultant or a counselor who can provide exactly the services you need. Ask for the references of people who have used the service. Know the qualifications of the counselor you will be working with. If you have any questions, check with the Better Business Bureau.

Listed here are several counselors/consultants who have established practices in the St. Louis area and/or are involved with professional organizations. The purpose of this book is not to endorse any firm or counselor/consultant, only to give an indication of the qualifications of some practitioners in the area. For a more complete listing, look in the Yellow Pages under *Vocational Guidance* and *Personnel Consultants*

Peg Atkins
747 N. Taylor
Kirkwood, MO 64122
(314) 822-3296
Career assessment and guidance, resume assistance and networking skills. Fee: Per session, sliding fee.

Career Planning Centers of America, Inc.
Thomas E. Brown, President
225 South Meramec, Suite 1226
Clayton, MO 63105
(314) 725-8122
Offers complete evaluation and counseling, including testing, resume preparation and interview skill development. Clients are taught to develop their own job referral networks. All professional staff members are nationally certified counselors. Fees: Hourly plus charge for testing with a sliding fee for some clients.

Careers UNLTD.
Lola Lucas
2211 S. Brentwood
St. Louis, MO 63144
(314) 727-5652
Offers career exploration and self-assessment. Assists individuals and groups in goal-setting and developing strategies to achieve their objectives. Some outplacement seminars. Fees: Hourly plus, sliding scale for some clients.

Joyce DeWoskin Psychological Services
12401 Olive St. Rd.
St. Louis, MO 63141
(314) 432-5845
Vocational testing and short- or long-term career counseling, career decision-making and

job-hunting skills. Licensed psychologist with fees depending on services selected.

Adele Levine
230 S. Bemiston
Suite 811
Clayton, MO 63105
(314) 725-2045
Specializes in career clarification or transition, teaches interviewing skills, job search techniques, redesigning resumes, balancing home and career. Fees: Per hour.

Missouri View Program
James H. Grogan, Ph.D., Director
15875 New Halls Ferry Road
Florissant, MO 63031
(314) 831-7100

Provides career counseling, resume assistance and testing. The Missouri View System is a statewide computerized career information delivery system listing current occupational and career decision-making information. (The program is available in all 114 Missouri counties at libraries and community colleges). Fees: First session free, then hourly fee.

Work Transitions
Anna Navarro
2256 S. 39th St.
St. Louis, MO 63110
(314) 773-1432
Individual consultation to determine what factors are important for one's work satisfaction, translating those factors into career goals and turning goals into reality. Clients trained to develop their own job referral resources. Fees: Depend on services needed.

Employment Agencies

Employment agencies match specific job openings to a candidate's skills and experience. Some agencies specialize in a particular occupational field, such as data processing or engineering, but most do not handle middle or upper management positions. Executive search firms or outplacement organizations (working with companies forced to reduce employees in a particular department) normally conduct searches for this group.

If you decide to select an agency, it is recommended that you read the agency's contract carefully before signing.

Consider these factors before deciding who can serve you best: the reputation of the agency, the type and level of job you seek, the companies the agency does business with, and the agency's success rate.

A listing here does not constitute an endorsement of a particular agency. Check with others who have used the service and consult with the Better Business Bureau (531-3300) if you have any questions. For a more complete list, check the Yellow Pages.

**ABC Management/
Employment Service**
25 S. Bemiston, Suite 214
Clayton, MO 63105
(314) 725-3140

For 25 years providing employment services to college graduates with job experience, particularly engineers. Employer pays placement fee.

Dynamic Employment Service
11500 Olive Blvd.
Suite 282
St. Louis, MO 63141
(314) 993-8883

Offering placement services in accounting, engineering, data processing, administrative, technical and sales positions. Fees: One percent per thousand of annual salary (example: for $20,000, 20 percent or $4,000 fee).

**Executive Career
Consultants, Inc.**
111 N. Taylor Ave.
St. Louis, MO 63122
(314) 965-3939

Specializes in data processing employment and recruiting. Fees assumed by client companies.

Robert Half of St. Louis, Inc.
7733 Forsyth Blvd.
Clayton, MO 63105
(314) 727-1535

National agency specializing in bookkeeping, accounting and data processing, both permanent and temporary placements. Employer pays fee.

**Professional Career
Development**
7777 Bonhomme, 11th Floor
Clayton, MO 63105
(314) 727-7670
One Mercantile Center, Suite 2610
St. Louis, MO 63101
(314) 436-5464

Clayton location offers professional services in banking, data processing and accounting. St. Louis office provides entry-level key-punch, secretarial, clerical and support personnel positions. Employer pays fee.

The Search Professionals
130 S. Bemiston Ave.
Suite 608
Clayton, MO 63105
(314) 727-1706

Nationally-affiliated group specializing in accounting, financial and medical fields. Employer pays fee.

J. Wottowa and Associates
910 N. 11th St.
Suite 200
St. Louis, MO 63101
(314) 621-4900
Recruiting in St. Louis since 1978 specializing in entry-level management, sales and mid- to upper-level insurance, banking and finance positions. Clients, not employer, pay the fee.

Gene Wanner Personnel
314 N. Broadway
St. Louis, MO 63102
(314) 621-8588
Specializes in administrative, sales, technical and clerical placements from trainee through executive level positions. Employer pays fee in some instances, others paid by client from 8 percent to 25 percent of first year's salary.

Executive Search and Outplacement Firms

These firms place people who have special qualifications with companies that have a specific employment need. Hiring companies employ headhunters, as executive searchers are often called, to search a wide area to locate a particular type of individual to meet the company's needs. Sometimes it is difficult to get an appointment with one of these companies unless you have the experience they seek. Normally outplacement firms work with corporations or industries that are reducing personnel because a product line, a subdivision or a management level is being eliminated. Mid- to upper-level employees who realize their jobs are being phased-out may be able to negotiate outplacement counseling as part of their termination agreement. If you find yourself in this position, it is important to negotiate this agreement while you are still employed by a corporation. Most outplacement firms do not work with individuals, rather they are paid by a particular company to locate positions for a group of employees who will be leaving a company.

Following is a list of selected executive and outplacement search firms. Look in the Yellow Pages for more information.

Career Dimensions
Judy Dubin
P. O. Box 8488
St. Louis, MO 63132
(314) 388-5158
Human resources company providing short- and long-term outplacement for individuals and groups; career development planning and personal development consulting; member of American Society for Training and Development. Fees paid by individual or corporate client.

Christopher and Long
15 Worthington
Maryland Hts, MO 63043
(314) 576-6300
Executive search firm specializing in national recruiting and place-ment of middle management-level personnel in data processing, banking, engineering, the agricultural business, liscensed physicians and surgeons. Employer pays the fee.

Drake Beam Morin Inc.
7701 Forsyth, Suite 1295
Clayton, MO 63105
(314) 725-7441
International outplacement firm locating executive-level employees for client companies. Services include workshops in interviewing, and programs to assist with spouse relocation counseling, directional counseling, for employees seeking change within a company, and pre-retirement

planning. Full-time psychologist on staff. Fees paid by company reducing management positions.

HDB, Incorporated
1302 Clarkson-Clayton Center
Suite 209B
Ellisville, MO 63011
(314) 391-7799

Large national executive recruiter specializing in data processing, artificial intelligence, defense-related industries, teleprocessing and telecommunications positions in the $23,000 to $150,000 range. Offers advice on interviewing techniques, attire, job research and resume follow-up. Graduating college seniors can take advantage of a $150 seminar providing resume assistance, interviewing skills and role-playing. Except for this seminar, fee paid by employer.

Right Associates
222 S. Central Ave., Suite 1000
Clayton, MO 63105
(314) 725-5700

Outplacement and human resource management consultants assisting with decision making, job search strategy, interviewing skills and resume editing. Exten-

sive experience in aerospace and business management gained after merger with Dice Cowger and Associates.

Search Financial
1515 N. Warson Rd.
Suite 145
St. Louis, MO 63132
(314) 423-8080

Executive search firm specializing in accounting. Employer pays fee.

G. M. Snyder and Associates
2190 S. Mason, Suite 305
St. Louis, MO 63131
(314) 821-9395

Recruiting firm specializing in national placements for technical positions for plant manager staff level and above in the areas of metals manufacturing, automation and robotics. Prospective candidates are invited to submit resumes. Employer pays fee.

Source Finance
111 West Port Plaza, Suite 810
St. Louis, MO 63146
(314) 434-7070

Recruiting and executive search firm specializing in the fields of accounting and finance. Employer pays fee.

Temporary Employment Agencies

Consider using temporary employment agencies for part-time placement if you need supplemental income while you are job hunting or want to develop new skills, contacts, and explore a particular field of interest.

Use this selected list of temporary employment agencies as a beginning, but be imaginative and look to create a part-time, temporary job in a company where you are doing informational interviewing. Temporary positions are not limited to clerical jobs, but are also available in marketing, health care, data processing and some professional areas. Several of the larger temporary agencies have more than one location, check the Yellow Pages for the office nearest you.

Accountemps of St. Louis
7733 Forsyth
Clayton, MO 63105
(314) 727-1535

Part-time positions for bookkeepers and accountants as well as clerks and secretaries.

B. Loehr Temporaries
1108 Olive St.
St. Louis, MO 63101
(314) 421-1688

Kelly Services
7777 Bonhomme, Suite 1505
Clayton, MO 63105
(314) 721-1995

LPL Technical Service Inc.
500 Northwest Plaza, Suite 905
St. Ann, MO 63074
(314) 739-9540
Temporary employment for engineers, designers and drafters.

Olsten Temporary Service
2222 Schuetz Rd., Suite 220
Creve Coeur, MO 63146
(314) 432-8088

Stivers Temporary Personnel
3270 Hampton Ave., Suite 2166
St. Louis, MO 63139
(314) 781-1900

Western Temporary Services
130 S. Bemiston Ave., Suite 609
Clayton, MO 63105
(314) 727-8845

ST. LOUIS METROPOLITAN AREA MAJOR EMPLOYERS

Despite a reputation built on beer and baseball, St. Louis takes corporate America seriously. Anheuser-Busch's empire extends to baked goods, amusement parks and snack foods, as well as its ever popular suds. But by no means does Busch reign supreme, rather it is one of many national and international companies that grew up here – companies like Ralston-Purina, McDonnell Douglas, Monsanto, Maritz and A.G. Edwards – and have remained to become anchors of the community.

St. Louis' location, near the center of the country, easily accessible by plane, train and the Missouri and Mississippi rivers, drew industries here before the turn of the century. The defense industry and aeronautics companies like McDonnell Douglas, one of the nation's largest aircraft producers, and General Dynamics, one of the largest shipbuilders, have combined sales/budgets approaching $30 billion annually. When the branches of the U.S. Air Force Command at Scott Air Force Base and the Army Systems Command are added to the defense contractors, approximately 60,000 people are employed by the defense industry in the Metropolitan area.

The diversity of industries established here during the last century required a variety of services, among them health care and educational opportunities for their employees. As a result 48 hospitals are scattered throughout the area and more than a dozen colleges. Among these institutions are a large community college system (St. Louis Community College), a major research university (Washington University), a large public university (University of Missouri-St. Louis), and one of the oldest Catholic universities in the nation (St. Louis University), plus a variety of smaller public and private colleges.

This selective listing of St. Louis area employers is to help you begin a survey of companies and organizations in fields that interest you. We have selected employers in each occupational category who employ the most people. Some larger employers are restructuring or reducing their headquarters personnel. But smaller companies, many located in suburban areas where overhead is lower, are often an excellent source of employment, particularly for less experienced workers or those re-entering the job force. For a more complete listing you can look in the Yellow Pages. Also check the resources listed in this guide titled, Directories and Publications, including *Sorkins Guide* available at most public libraries.

Accounting Firms

Changes in the federal tax code for both businesses and individuals have created a healthy environment for accountants – particularly smart ones who are keeping pace. Of course there is a lot more to accounting than figuring out what's owed to Uncle Sam. A variety of firms are available to choose from, whether you're looking for a corporate environment or a small firm catering to a diverse clientel.

Arthur Anderson & Co.
1010 Market St.
St. Louis, MO 63101
(314) 621-6767

Coopers & Lybrand
One Mercantile Center
St. Louis, MO 63101
(314) 436-3200

Ernst & Whinney
Gateway One, Suite 1400
701 Market St.
St. Louis, MO 63101
(314) 231-7700

Grant Thornton
500 Washington
Suite 1200
St. Louis, MO 63101
(314) 241-8881

Price Waterhouse
One Centerre Plaza
St. Louis, MO 63101
(314) 425-0500

Rubin, Brown, Gornstein & Co.
230 South Bemiston Ave.
Clayton, MO 63105
(314) 727-8150

Touche, Ross & Co.
2100 Railway Exchange
St. Louis, MO 63101
(314) 231-3110

Advertising Agencies

Local ad firms represent a broad range of clients stretching from the corner grocer to multi-national corporations, so there are ample opportunities for someone with imagination and no adversion to hard work. Thanks to changing technology, St. Louis firms are doing many of the jobs that once required a call to Chicago or New York. The larger agencies have offices in major cities and provide opportunities to interact with industry leaders.

BHN Advertising & Public Relations
910 North 11th St.
St. Louis, MO 63101
(314) 241-1200

D'Arcy Masius Benton & Bowles
One Memorial Drive
St. Louis, MO 63102
(314) 342-8600

Gardner Advertising Co.
10 Broadway
St. Louis, MO 63102
(314) 444-2000

**Frank James Direct
Marketing Co.**
120 S. Central Ave., Suite 500
Clayton, MO 63105
(314) 726-4600

Kelly Inc.
2850 S. Jefferson Ave.
St. Louis, MO 63118
(314) 646-0023

*Markets nationally with complete
mail marketing service.*

**Kenrick Advertising Inc.
(Sub. of Aragon Co.)**
7711 Carondelet Ave.
Clayton, MO 63105
(314) 726-6020

McGavren-Guild Radio Inc.
10 S. Broadway
Suite 472
St. Louis, MO 63102
(314) 231-0000

*National rep firm selling radio
time.*

Stolz Advertising Co.
7701 Forsyth Blvd.
Clayton, MO 63105
(314) 863-0005

Talent Plus
3663 Lindell Blvd.
Suite 100
St. Louis, MO 63108
(314) 531-4800

*Provides personalities and mod-
els for TV and radio locally and
nationally.*

Architectural Firms

Rated by *U.S. News and World Report* as the nation's leader in historic preservation, St. Louis surpassed other strong contenders like New Orleans, Philadelphia and Boston, spending nearly $350 million to renovate more than 600 historic properties between 1982 and 1985. Many of the city's leading architects fought the battles to restore these aging classics and have reaped the rewards. Helmuth Obata & Kassabaum, Inc. undertook one of the most complex projects — breathing life into a 90-year-old train station. Today Union Station features unusual one-of-a-kind shops and nearly every culinary delight from homemade fudge to fresh seafood. An Omni Hotel anchors one edge of the station, overlooking a beer garden and a pond built within the 11.5-acre train shed. Shoppers are also drawn to St. Louis Centre (RTKL Associates), one of the largest urban shopping malls in the country. Clayton, at one time St. Louis' only suburban business center, is experiencing a building boom, but has competition from other business districts, notably Chesterfield on the western edge of St. Louis County, St. Charles in the northwest and several other locations both north and south of the city.

**(Kenneth) Balk &
Associates Inc.**
1066 Executive Parkway
St. Louis, MO 63141
(314) 576-2021

Robert L. Boland, Inc.
12935 N. Forty Dr.
St. Louis, MO 63141
(314) 878-9205

Booker Associates, Inc.
1139 Olive St.
St. Louis, MO 63101
(314) 421-1476

*Operates internationally, 28
architects, 89 engineers.*

HBE Corporation
11330 Olive St. Rd.
Creve Coeur, MO 63141
(314) 567-9000

**Hellmuth Obata &
Kassabaum, Inc.**
100 N. Broadway
St. Louis, MO 63102
(314) 421-2000

*(Union Station, Edison Brothers
Headquarters, St. Louis Galleria)*

**Peckham, Guyton,
Albers & Viets Inc.**
200 N. Broadway
St. Louis, MO 63102
(314) 231-7318

**Louis R. Saur &
Associates, Inc.**
168 N. Meramec Ave.
Clayton, MO 63105
(314) 727-4484

Sverdrup Corporation
801 N. Eleventh St.
St. Louis, MO 63101
(314) 436-7600

*(Lambert Field Concourse D and
International Wing)*

Aeronautics and Defense Contractors

St. Louis' strong ties to the aeronautics industry began more than 40 years ago with the opening of McDonnell Douglas Corporation, now the largest employer in the metropolitan area. The growth of McDonnell Douglas and its crosstown rival, General Dynamics, has fostered the development of the defense industry here. Scott Air Force Base in Illinois houses the U.S. Air Force Military Airlift Command, and Communications Command, while the Army Troop Support Command and the Air Force Mapping Agency are located in St. Louis. McDonnell Douglas is a major producer of the Tomahawk cruise missile and Harpoon anti-ship missile, as well as the F-15 jet and major portions of the Space Shuttle. General Dynamics produces submarines, tanks, aircraft and strategic systems for processing and storing military computer data.

The defense industry also provides work for a variety of other major contractors and manufacturers in the metropolitan area. Listed below are a few of the largest local defense contractors.

MISSOURI

Emerson Electric Co.
8000 W. Florissant Ave.
St. Louis, MO 63136
(314) 553-2000

Radar systems, missile launching and guidance systems, airborne fire-control systems and anti-submarine systems.

Engineered Air Systems Inc.
1270 N. Price Road
Clayton, MO 63132
(314) 993-5880

General Dynamics
7733 Forsyth
Clayton, MO 63105
(314) 889-8200

International defense contractor producing ships, aircraft, submarines, tanks and strategic

systems for processing and storing computer data.

Valentec Kisco Co.
6300 St. Louis Ave.
St. Louis, MO 63121
(314) 381-9850

Containers for Sparrow and Sidewinder missles, Howitzer and tank guns.

McDonnell Douglas Corp.
McDonnell Blvd. and Airport Road
P.O. Box 516
St. Louis, MO 63166
(314) 232-0232

St. Louis' largest employer, now expanding into communications (see electronics/computers).

**U.S. Army Troop
Support Command**
4300 Goodfellow Blvd.
St. Louis, MO 63120-1798
(314) 263-2201
*Army warehouse and distribution
(TROSCOM) service.*

ILLINOIS

**U.S. Air Force
Communications Command**
Scott Air Force Base, IL 62225-6001
(618) 256-2571

**U.S. Air Force Military
Airlift Command**
Scott Air Force Base, IL 62225-5001
(618) 256-3205
Responsible for rapid airlift mobility and logistical air resupply of American forces.

Olin Corp.
427 N. Shamrock St.
East Alton, IL 62024-1174
(618) 258-2000
Leading national producer of ammunition .

Automotive

Missouri has the mixed blessing of being the second largest automotive producer in the country, producing over $3 billion in automotive products in St. Louis alone. Because of the continuing changes in the automotive industry some of the information below may be out-dated by the time of printing, so contact the local labor union or the local companies for more current information.

American Transit Corp.
120 S. Central Ave.
St. Louis, MO 63105
(314) 726-9200
Subsidiary of Chromalloy American Corp., second largest management consulting firm for transportation operations in the United States. Leading independent U.S. distributor of inter-city buses.

**Chrysler St. Louis Car
Assembly Plant I & II**
1001 N. Highway Dr.
Fenton, MO 63026-1997
(314) 343-2500
After producing the last Chrysler Fifth Avenue late in 1986, Plant No. 2 closed for 16 weeks to retool for the Dodge Caravan and Plymouth Voyager minivans. Plant No. 1 reopened after a $210 million expansion and retooling.

**Ford Motor Co.
St. Louis Assembly Plant**
6250 N. Lindbergh
Hazelwood, MO 63042
(314) 731-6300
Assembles Ford Aerostar vans.

**General Motors B-O-C
Wentzville Assembly Center**
1500 E. Route A
Wentzville, MO 63385
(314) 327-5711
One of GM's largest auto assembly plants, expected to be a mainstay of the "new" streamlined GM, currently produces Buicks and Oldsmobiles.

**Southwest Mobile
Systems Corp.**
200 Sidney St.
St. Louis, MO 63104
(314) 771-3950
Manufactures truck bodies, trailers and mobile maintenance shops for the military.

Banks

Legislation allowing banks to serve wider areas has increased competition and given an advantage to larger institutions. Again, as with the automotive industry, rapid changes in banking make it difficult to provide up-to-the-minute information. This listing includes only some of the larger institutions and is not meant to be all inclusive. Consult the *St. Louis Business Journal* for more current information on individual companies.

Boatmen's Bancshares Inc.
100 N. Broadway
St. Louis, MO 63102
(314) 425-7500

Centerre Bancorporation
One Centerre Plaza
St. Louis, MO 63101
(314) 554-6000

Commerce Bancshares
922 Walnut
Kansas City, MO 64141
(816) 234-2000
(800) 892-7100

Boatmen's National Bank of St. Louis
720 Olive St.
St. Louis, MO 63101
(314) 241-3600

Mark Twain Bancshares Inc.
8820 Ladue Road
St. Louis, MO 63124
(314) 727-1000

Savings and Loans

Community Federal Savings and Loan Association
One Community Federal Center
St. Louis, MO 63131
(314) 822-5000

Missouri Savings Association
10 N. Hanley Road
St. Louis, MO 63105
(314) 862-3300

Roosevelt Federal Savings and Loan Association
900 Roosevelt Parkway
Chesterfield, MO 63017
(314) 532-6200

First Nationwide Bank
8020 Forsyth Blvd.
Clayton, MO 63105
(314) 726-2800

CLAYTON

Clayton Savings and Loan Association
135 N. Meramec Ave.
Clayton, MO 63105
(314) 862-6900

ILLINOIS

Germania F. A.
543 East Broadway
Alton, IL 62002
(618) 465-5543
(314) 355-0700

Illini Federal Savings and Loan Association
6550 N. Illinois St.
Fairview Hts, IL 62208
(314) 241-5300

Chemical and Allied Products

In 1901 a young St. Louis chemist opened a plant to manufacture vanilla and other flavorings. Today the community around the original plant bears his name, Queeny, and the company bearing his wife's maiden name, Monsanto, is an international corporation with $7 billion annual sales. Despite several staff reductions at its Creve Coeur headquarters, Monsanto ranks among the top 10 employers in St. Louis. The company's chemists developed the herbicides *Lasso* and *Roundup*, as well as fibers for *AstroTurf* and *WearDated* carpet and textiles. Monsanto branched out in 1985 by acquiring the G. D. Searle Co., the world's 26th largest pharmaceutical firm and the producer of *NutraSweet*. More importantly, the pharmaceutical company gives Monsanto a chance to develop products based on its biotechnical research. Monsanto's 1986 research and development budget topped $450 million to aid in plant and microbial genetic engineering, human health care, animal science, cell culture, fermentaion, plant tissue culture and molecular biology. As part of this program, the company has a $62 million joint research project with Washington University being conducted through 1987.

St. Louis' other homegrown chemical company, Mallinckrodt, Inc., produces flavors, fragrances and the ingredient in pain-killers not containing asprin. In 1982 Avon acquired Mallinckrodt, then sold the company to International Minerals & Chemicals Corp. in 1986. Other local chemical companies are generally involved with the oil industry, which is retrenching due to fluctuations in the price of oil.

ST. LOUIS

Chemtech Industries, Inc.
1655 Des Peres Rd., P.O. Box 31000
St. Louis, MO 63131
(314) 966-9900

E-Z Clor System
1920 Beltway Dr.
St. Louis, MO 63114
(314) 426-4533

Ethyl Petroleum Corp.
20 S. Fourth St.
St. Louis, MO 63102
(314) 421-3930

Mallinckrodt Inc.
675 McDonnell Blvd.
St. Louis, MO 63042
(314) 895-2000

Monsanto Corp.
800 N. Lindbergh Blvd.
St. Louis, MO 63167
(314) 694-1000

Petrolite Corp.
100 N. Broadway
St. Louis, MO 63102
(314) 241-8370

ILLINOIS

Kaiser/Estech
Div. of Vigoro Industries, Inc.
2007 W. Highway 50
Fairview Hts, IL 62208
(618) 624-5522

Communications

Deregulation of the industry has provided unusual opportunities for smaller companies and ingenious people with new ideas.
However, larger firms like AT & T, Southwestern Bell and MCI have
trimmed management-level employees to remain competitive. Even
with these cuts in personnel, Southwestern Bell is still expected to
have nearly 60,000 workers here in 1988, while AT & T personnel
has remained at about 4,500. MCI plans to make staff reductions
to remain competitive with its larger competitors.

MCI Southwest has a brand-new building downtown and a sales
and customer service branch near West Port. In its bid to outdistance
SPRINT and other competitors in the long-distance market, MCI
signed a contract to install its own fiber optic cables.

Companies like LDX Net, Inc., headquartered in Chesterfield,
are where the greatest long-term growth is expected. Using access
along rail lines of its former parent, Southern (Rail) Industries of
Kansas City, LDX began providing voice and data networks to cities
in the same five-state area served by SW Bell and MCI's Southwestern
Division located here.

American Telephone & Telegraph Communications
424 S. Woods Mill Rd.
Chesterfield, MO 63017
(314) 275-3000
*Feeling the pinch from new unre-
gulated long-distance and data
transmitters.*

MCI Telecommunications
One Centerre Plaza, Ste 1500
St. Louis, MO 63101
(314) 342-7300

Regional Office:
100 S. Fourth St.
St. Louis, MO 63102
*Coordinates sales, marketing,
operations, finance and administration for a five-state region.*

Southwestern Bell
One Bell Center
St. Louis, MO 63101
(314) 235-9800
*One of the largest communica-
tions services in the country
serving Missouri, Arkansas,
Kansas, Oklahoma and Texas.*

Cybertel Corp.
100 Ludwig Dr.
P.O. Box 4425
Fairview Hts, IL 62208
(314) 444-4444
*Cellular telephone and mobile
phone service; oldest and largest
of its kind in Missouri and Illinois,
outside Chicago.*

Computers

Although St. Louis hasn't made headlines as a haven for computer wizards, a glance at Sunday's *Post Dispatch* employment section indicates a strong market still exists here. Computers and people
who know how to use them are important to St. Louis partly because
of the role technology plays in the defense industry and among the
major corporations headquartered there. Likewise the telecommunications industry, with large employers like Southwestern Bell and
MCI, also relies heavily on computer technology.

AT & T Network Systems
1111 Woods Mill Rd.
Ballwin, MO 63011
(314) 391-2000

*Serves western half of the U.S.
providing engineering and support services.*

**Digital Equipment
Corporation (DEC)**
721 Emerson Rd.
P.O. Box 27320
St. Louis, MO 63141
(314) 991-6400

*Second to IBM in the local market
for business use.*

Hewlett-Packard
13001 Hollenberg
Bridgeton, MO 63044
(314) 344-5100

**International Business
Machines**
500 Maryville College Dr.
St. Louis, MO 63141
(314) 469-4000

**McDonnell Douglas
Information Systems Group
(MDISG)**
P.O. Box 516
St. Louis, MO 63166
(314) 232-0232

*One of the largest data processing
companies in the U.S., serving 1
of 5 hospitals in the country.*

**Monsanto Engineered
Products Division**
P.O. Box 8
501 Pearl Dr.
St. Peters, MO 63376
(314) 272-6281

*Manufactures electronic-grade
silicon wafers for use in manufacture of integrated circuits.*

Olin Brass Group
Shamrock St.
East Alton, IL 62024
(618) 258-2000

*Largest U.S. producer of lead
frames for computer chips.*

Construction

Despite an abundance of new construction work in the St. Louis
area, trade unions are making concessions to keep the cost of construction down and maintain the union market here. Local contractors are not required to hire union workers, nevertheless union
employees do about 90 percent of construction work in St. Louis,
compared to 30 percent nationwide.

Alberici Construction
2150 Kienlen Ave.
St. Louis, MO 63121
(314) 261-2611

*Major national contractor.
(Crestwood Plaza renovation, GM
Wentzville assembly plant)*

Fru-Con Corp.
1299 W. Clayton Rd.
Ballwin, MO 63011
(314) 391-6700

(Busch Stadium, Centerre Plaza)

HBE Corp.
11330 Olive Blvd.
St. Louis, MO 63141
(314) 567-9000

*(Adams Mark Hotel, Alexian
Brothers Hospital)*

Hercules Construction Co.
8220 Delmar Blvd.
St. Louis, MO 63124
(314) 991-3730

*(St. Mary's Health Center, Apex Oil
Complex).*

Ralph Korte Construction Co.
700 St. Louis Union Station
St. Louis, MO 63103
(314) 231-3700

*(Linclay Corp., Union Station
annex)*

McCarthy Brothers Co.
1341 N. Rock Hill Rd.
St. Louis, MO 63124
(314) 968-3300

*(One Gateway Mall, Edison
Brothers Bldg.)*

C. Rallo Contracting Co.
5000 Kemper Ave.
St. Louis, MO 63139
(314) 664-2900
(Lambert Airport, Convention Center)

Fred Weber Inc.
7929 Alabama Ave.
St. Louis, MO 63111
(314) 638-1570
(Washington University School of Business, Arch Parking Garage)

Educational Institutions

The metropolitan area has colleges and universities of nearly every size, religious persuasion and academic specialty. Career counseling centers are available at most of these institutions. Some offer assistance with internships and on-site job interviews.

MISSOURI

Concordia Seminary
801 DeMun Ave.
St. Louis, MO 63105
(314) 721-5934
Luthern theological seminary.

Fontbonne College
6800 Wydown Blvd.
St. Louis, MO 63105
(314) 862-3456
Private four-year college formerly administered by Catholic nuns. Known for speech and hearing, special ed training, home economics and small fine arts courses.

Harris Stowe State College
3026 Laclede Ave.
St. Louis, MO 63103
(314) 533-3366
Four-year state college, primarily a teacher's college with emphasis on urban education.

Lindenwood College
Kingshighway and First Capitol Dr.
St. Charles, MO 63301
(314) 946-6912
Private college offering BA in computer science, arts and sciences, plus MBA and MA in education.

Maryville College
13550 Conway Rd.
St. Louis, MO 63141
(314) 576-9300
Private college with more students attending the evening MBA program part-time than attending full time during the day.

Missouri Baptist College
12542 Conway Road
St. Louis, MO 63141
(314) 434-1115

St. Louis College of Pharmacy
Euclid Avenue and Parkview Place
St. Louis, MO 63110
(314) 367-8700

St. Louis Community College
5801 Wilson Avenue
St. Louis, MO 63110
(314) 644-9550
Large two-year community college with campuses throughout the metropolitan area.

St. Louis University
221 N. Grand Blvd.
St. Louis, MO 63101
(314) 658-2222
Private Jesuit university founded in 1818 offering 34 different degrees, including doctorates in law, medicine, psychology, arts and sciences.

University of Missouri-St. Louis
8001 Natural Bridge Rd.
St. Louis, MO 63121
(314) 553-5000
Large public university offering 19 degree programs.

Washington University
One Brookings Drive
St. Louis, MO 63130
(314) 889-5000
Private research university offering 41 degree programs, including doctorates in law, dentistry, medicine, arts and sciences. Among the top five endowed universities in the country.

Webster University
470 E. Lockwood
St. Louis, MO 63119-3194
(314) 968-6900
Private university known for its programs in the arts, offering degrees at both the BA and MA levels.

ILLINOIS

Belleville Area College
2500 Carlyle Rd.
Belleville, IL 62221
(618) 235-2700
Two-year community college.

McKendree College
701 College Rd.
Lebanon, IL 62254
(618) 537-4481
Four-year liberal arts college with BS in nursing.

State Community College
601 James R. Thompson Blvd.
East St. Louis, IL 62201
(618) 274-6666
Two-year community college offering associate's degrees in arts, sciences and applied sciences.

Southern Illinois University-Edwardsville
Edwardsville, IL 62026
(618) 692-2000
(314) 621-5168
Public university offering 35 different degrees, including doctorates in dentistry and education.

Electronics

St. Louis' close ties to the defense industry, which relies heavily on developments in electronics, provide the base for this industry locally. The instability of contract work in the defense and aerospace industry has encouraged some firms to diversify. In some cases moving into other areas has been successful, not in others.

Bussman - Div. of Cooper Industries
114 Old State Road
P.O. Box 14460
St. Louis, MO 63178
(314) 394-2877
Major manufacturer of fuses and electrical equipment.

Emerson Electric Co.
8000 W. Florissant Ave.
St. Louis, MO 63136
(314) 553-2000
Major international manufacturer of commercial electronics components, appliances and tools.

Fisher Controls International
8000 Maryland Ave.
Clayton, MO 63105
(314) 694-9900
Largest manufacture of automatic control valves.

General Dynamics
7733 Forsyth Blvd.
St. Louis, MO 63105
(314) 889-8200
Data systems developed primarily for military purposes.

Engineering

An indication of the diversity of engineering positions in St. Louis is the Engineer's Club, an umbrella group for 52 separate organizations. Listed here are some of the largest engineering firms, but many area engineers are employed by major firms like McDonnell Douglas and Monsanto.

Balk & Associates Inc.
14500 S. Outer Rd.
Chesterfield, MO 63107
(314) 576-2021
Full service engineering and architectural company.

Booker Associates Inc.
1139 Olive St.
St. Louis, MO 63101
(314) 421-1476
International engineering firm.

Fruco Engineers
1299 Clayton Rd. W.
Ballwin, MO 63011
(314) 391-8866
Full-service process, civil, structural engineers operating internationally.

Marston & Marston Inc.
1655 Des Peres Rd.
Suite 150
Des Peres, MO 63131
(314) 822-2254
Specializes in mining engineering, contract mining of coal and precious metals worldwide.

Sheppard Morgan & Schwaab Inc.
215 Market St.
Alton, IL 62002
(618) 462-9755
One of oldest firms in Metro-East (1900), specializes in transportation, environmental engineering.

Sverdrup Corp.
801 N. Eleventh St.
St. Louis, MO 63101
(314) 436-7600
Developed by a St. Louis transplant in 1928, the company today employs more than 1,000 engineers throughout the world.

Tao (William) & Associates, Inc.
2357 59th St.
St. Louis, MO 63110
(314) 644-1400
Diversified consulting civil, structural, electrical and mechanical engineering firm with computer services subdivision.

Telcom Services Inc.
18040 Edison Ave.
Chesterfield, MO 63017
(314) 532-3535
Communications engineers, electronic installation and testing, systems programming.

Entertainment and Culture

The World's Fair of 1904 put St. Louis on the map. Although only one building from the Fair remains (the Museum of Art in Forest Park), the event established St. Louis' cultural standards. It brought an influx of talented people with new ideas and potent dreams. Today sons and daughters of those Fair visitors who came back to stay are supporting their parents' legacy in everything from the Gateway Arch and the Jefferson Memorial to the Historical Society and the St. Louis Zoo, one of the nation's finest.

AMC Theaters
6636 Clayton Rd.
St. Louis, MO 63117
(314) 781-5000

Bach Society of St. Louis
470 E. Lockwood Ave.
Webster Groves, MO 63119
(314) 962-0669

Busch Entertainment Corp.
One Busch Place
St. Louis, MO 63118
(314) 577-4715

Operates theme parks (Busch Gardens) throughout the country in Florida, Virginia and Pennsylvania.

CASA
(St. Louis Conservatory of Music)
560 Trinity Ave.
University City, MO 63130
(314) 863-3033

Dance St. Louis
149 Edgar Rd.
Webster Groves, MO 63119
(314) 968-3770

Local troupe also sponsors visits by major dance companies at Kiel Auditorium.

Edison Theater
Mallinckrodt Center
Washington University
P.O. Box 1119
St. Louis, MO 63105
(314) 889-6518

Offers a stage for visiting performers like John Houseman's Acting Company, Minneapolis' Guthrie Theater, as well as local and national dance companies.

Jefferson National Expansion Museum
11 N. 4th St.
St. Louis, MO 63102
(314) 425-4465

More than 3.5 million visitors come each year to see the tallest national monument, the symbol of St. Louis' role as the Gateway to the West.

Metropolitan Zoological Park and Museum
135 N. Meramec Avenue
Forest Park
St. Louis, MO 63105
(314) 862-4222

Known as one of the nation's outstanding zoos, recent renovation included a new Ape House.

Opera Theater of St. Louis
1 Kirthom Lane
P.O. Box 13148
Webster Groves, MO 63119
(314) 961-0171

A stage for local talent in cooperation with Webster College.

Missouri Historical Society
Jefferson Memorial
Forest Park
St. Louis, MO 63112
(314) 361-1424

The Historical Society built its imposing structure at the main gate site of the 1904 Exposition from the Fair's surplus funds. Inside the museum preserves the stories of the trappers, steamboat captains, military heroes, brewmasters, pilots, baseball players, industrial and religious leaders who built St. Louis and helped it grow.

The MUNY
Municipal Theatre Association of St. Louis
Forest Park
St. Louis, MO 63112
(314) 361-1900

Billed as the world's largest outdoor stage, the MUNY opened in 1919, presents a variety of musicals and popular-media performers.

Repertory Theatre of St. Louis
130 Edgar Rd.
Webster Groves, MO 63119
(314) 968-7340

Professional performances in conjunction with Webster College showcase local talent.

St. Louis National Baseball Club Inc.
250 Stadium Plaza
St. Louis, MO 63102
(314) 421-3060

Baseball's been a tradition in St. Louis since 1875, when 10,000 watched the Brown Stockings beat the Chicago White Stockings 10-0. A perennial contender in the National League, the Cardinals have some of the most loyal fans in the country.

St. Louis Black Repertory Co.
2240 St. Louis Ave.
St. Louis, MO 63106
(314) 231-3706

A stepping stone for talented local black performers.

St. Louis Art Museum
Forest Park
St. Louis, MO 63110
(314) 721-0067

Designed by famed architect Cass Gilbert in 1904, the museum first served as the Fine Arts Palace of the World's Fair. Attendance averages three-quarters of a million visitors a year who come to see a fine collection of primitive art, 19th century French Impressionist paintings, works by Missourian George Caleb Bingham, and a section devoted to modern painting and sculpture.

St. Louis Blues Hockey Club
5700 Oakland Ave.
St. Louis, MO 63110
(314) 781-5300

Sale of the Blues to a local investor insures hockey will be in St. Louis for years to come.

St. Louis Football Cardinals
200 Stadium Plaza
P.O. Box 888
St. Louis, MO 63102
(314) 421-0777

Football came to town in 1960 when the Chicago Cardinals moved west. Support for the team grew after 1965 when Busch Stadium opened.

St. Louis Science Center
Forest Park
5100 Clayton Ave.
St. Louis, MO 63110
(314) 289-4400

A unique experience for children and adults alike.

St. Louis Steamers
212 N. Kirkwood Rd.
St. Louis, MO 63122
(314) 821-1111

St. Louis University reigned as the source of soccer talent for decades, but the area junior colleges are also producing good players. The Steamers have taken advantage of the popularity of the support.

St. Louis Symphony
Powell Hall
718 N. Grand Blvd.
St. Louis, MO 63103
(314) 533-2500

Rated by many critics as the best symphony in the country, the St. Louis Symphony delights audiences from London to Tokoyo.

Wehrenberg Theaters
1215 Des Peres Rd.
St. Louis, MO 63131
(314) 822-4520

Six Flags Over Mid-America
I-44 & Allenton Six Flags Rd.
Eureka, MO 63025
(314) 938-5300

St. Louis area's major amusement park.

Food & Beverages

Blessed with plenty of water from the Mississippi and abundant grain resources nearby, St. Louis provides a natural setting for the brewing and baking industries. Before the Civil War, 40 local breweries produced $1.76 million in sales. By 1900 only 19 survived with Anheuser-Busch pulling ahead as the largest brewer in the country. Today Busch has expanded into bakery and snack foods, wine and the operation of the St. Louis Baseball Cardinals. A low calorie wine cooler, *Dewey Stevens Premium Light*, breaks new ground in this competitive market. With over $6.5 billion in sales, Busch is one of the larger employers in St. Louis.

Like the breweries, the food distribution and manufacturing firms initially depended on Midwestern farmers and convenient transportation by river and rail. Western wheat, praised for its low moisture content, helped establish the reputation of St. Louis' flour and led to the opening of the first miller's exchange here in 1850.

These basic industries laid the foundation for St. Louis' role as a food production and distribution center. Ralston Purina, also one of the city's larger employers, has nearly one-third of the pet food market.

Anheuser-Busch Co. Inc.
One Busch Place
St. Louis, MO 63118
(314) 577-2000
One of the largest brewers in the country.

Coca-Cola Bottling Co. of St. Louis
19 Worthington Dr.
Maryland Heights, MO 63043
(314) 878-0800

ConAgra, Inc.
145 W. Broadway
Alton, IL 62002
(618) 463-4411
Processes grain into flour, feed and feed pellets.

Leaf, Inc.
(formerly Switzer Clark Candy Co.)
1600 N. Broadway
St. Louis, MO 63102
(314) 421-3474
Acquired Milk Duds, Good & Plenty and Now Later in 1983.

Pillsbury Company Plant
5020 Shreve Ave.
St. Louis, MO 63115
(314) 385-9100
Manufactures floor and feed.

Pet Inc.
Pet Plaza
400 S. Fourth St.
St. Louis, MO 63102
(314) 621-5400
Subsidiary of IC Industries; Old El Paso Mexican food, Whitman's Chocolates, Pet-Ritz pie shells.

Ralston-Purina Co.
Checkerboard Square
835 S. 8th St.
St. Louis, MO 63164
(314) 982-1000
Largest producer of dry dog and cat food, manufactures fresh bakery products and acquired Eveready/Energizer battery in 1986. Subsidiary Continental Baking Co. produces cakes and snacks at 48 sites.

Schnuck Markets
12921 Enterprise Way
Bridgeton, MO 63044
(314) 344-9600
Largest grocery retailer in area.

Wetterau Incorp.
8920 Pershall Rd.
Hazelwood, MO 63042
(314) 524-5000
Large grocery retailer, serving 26 states.

Furniture & Appliances

Foreign competition is forcing American furniture manufacturers to trim expenses, lay off employees and to develop new markets (like the company that manufactures kitchen cabinets economically using a vinyl veneer that resembles wood). Despite stiff competition from warehouse-style companies, several family-run stores remain in St. Louis, possibly because there are still many people who have lived in the metropolitan area for generations. They return to purchase their stoves, couches and washing machines from the descendants of the people their parents and grandparents trusted when they set up housekeeping.

Bank Bldg. & Equipt Corp.
1130 Hampton Ave.
P.O. Box 5137
St. Louis, MO 63139
(314) 647-3800
Designs and makes furniture for banks, hospitals and industry. Employees are considering purchase of firm to ward off unfriendly take-over.

Blackburn-(Division of FL Industries)
1525 Woodson Rd.
St. Louis, MO 63114
(314) 993-9430

Manufactures lighting fixtures and electrical products

Carafiol Furniture Co.
12100 St. Charles Rock Rd.
Bridgeton, MO 63044
(314) 291-6330

Charleswood Furniture Corp.
15400 S. Outer 40 Rd.
Chesterfield, MO 63017
(314) 532-3344

Harvard Interiors
4321 Semple Ave.
St. Louis, MO 63120
(314) 382-5590

Commercial furniture manufacturer with electronics components division in Arnold, MO.

Kessler Furniture Co.
419 E. Gano St.
St. Louis, MO 63147
(314) 231-2700

Pat Riley Warehouse
12567 Natural Bridge
Bridgeton, MO 63044
(314) 291-3025

Orchard Corp. of America
1154 Reco Ave.
St. Louis, MO 63126
(314) 822-3880
Prints wood grain designs used to panel RVs and cover furniture

Rothman Furniture Stores
2101 E. Terra Ln
O'Fallon, MO 63366
(314) 928-3494

St. Louis Electric Supply
6801 Hoffman
St. Louis, MO 63139
(314) 645-9000

Schweig-Engel Co.
3458 N. Union Blvd.
St. Louis, MO 63115
(314) 381-2000

Slyman Brothers Appliances
11448 St. Charles Rock Rd.
Bridgeton, MO 63044
(314) 291-5522

Tipton Centers
969 Anglum Dr.
Hazelwood, MO 63042
(314) 731-6161

Townhouse Penthouse Industries
1314 Hanley Industries Ct.
St. Louis, MO 63144
(314) 968-1818
Manufacturer of upholstered wood furniture.

Welsh Industries
1535 S. 8th St.
St. Louis, MO 63104
(314) 231-8822
Baby carriage and furniture manufacturer since 1929.

Government

Political disagreements between the Democratic city organization and the Republican suburban government are mellowing in light of economic reality. Both governments need to present a united front to businesses and industries they want to lure to the metropolitan area. For example, officials worked together to insure that the Blues Hockey team, which currently skates in the old Arena, stays in St. Louis.

ST. LOUIS

City Hall
1200 Market St.
St. Louis, MO 63101
(314) 622-3201

Agency on Training & Employment
317 N. 11th St.
St. Louis 63101
(314) 241-4300

Health & Hospitals Dept.
634 N. Grand
St. Louis, MO 63101
(314) 658-1140

Metropolitan Police Department
1200 Clark St.
St. Louis, MO 63103
(314) 231-1212

Parks & Recreation Dept.
5600 Clayton Rd.
St. Louis, MO 63110
(314) 535-5050

St. Louis Public Schools
911 Locust St.
St. Louis, MO 63101
(314) 231-3720

Dept. of Public Utilities
1220 Carr Lane
St. Louis, MO 63104
(314) 664-8070

Transportation & Traffic Division
1900 Hampton Ave.
St. Louis, MO 63139
(314) 647-3111

CLAYTON

City Government
10 N.Bemiston
Clayton, MO 63105
(314) 727-8100

ST. LOUIS COUNTY

County Executive
41 S. Central Ave.
Clayton, MO 63105
(314) 889-2000

County Dept. of Human Resources
555 S. Brentwood Blvd.
Clayton, MO 63105
(314) 889-3453

Co. Parks & Recreation
41 S. Central Ave.
St. Louis, MO 63105
(314) 889-2894

Co. Police Dept.
7900 Forsyth Blvd.
St. Louis, MO 63105
(314) 889-2260

Dept. of Public Works
7900 Forsyth Blvd.
Clayton, MO 63105
(314) 889-2484

Cooperating School Districts of St. Louis Suburban Area
1460 Craig Rd.
St. Louis, MO 63146
(314) 872-8282
Association of 45 area school districts

ILLINOIS

East St. Louis City Government
301 Broadway
East St. Louis, IL 62204
(618) 482-6811

St. Clair County
10 Public Square
Belleville, IL 62220
(618) 277-6600

Madison County
155 N. Main
Edwardsville, IL 62025
(618) 692-6290

Federal Government

Dept. of Agriculture
Extension Office
8001 Natural Bridge Rd.
St. Louis, MO 63121
(314) 553-5184

Economic Development Administration
210 N. Tucker
St. Louis, MO 63101
(314) 425-4312

Defense Logistics Agency (Contracts Administration)
1136 Washington Ave.
St. Louis, MO 63103
(314) 263-6510

Federal Bureau of Investigation
1520 Market
St. Louis, MO 63103
(314) 241-5357

Food & Drug Administration
808 N. Collins St.
St. Louis, MO 63103
(314) 425-5021

Internal Revenue Service Federal Tax Info
1114 Market St.
St. Louis, Mo 63103
(314) 342-1040

Social Security Administration
219 S. Central Ave.
St. Louis, MO 63105
(314) 679-7800

Hospitals

With 48 hospitals in the Metropolitan St. Louis area, nurses and other health care professionals have numerous opportunities and can find financial rewards, particularly in specialty areas. Competition for patients has increased significantly in recent years. New cost-cutting restrictions in health care programs by some of the larger local employers have also pressured area hospitals and physicians. Listed below are some of the larger hospitals; some are also selected for location.

Barnes Hospital
4949 Barnes Hospital Plaza
St. Louis, MO 63110
(314) 362-5000
Research hospital, largest in area with 1,200 beds; affiliated with Washington University.

Christian Hospital Northeast
11133 Dunn Rd.
St. Louis, MO 63136
(314) 355-3800

Northwest: 1225 Graham Rd.
Florissant, MO 63031
(314) 839-3800

DePaul Health Center
12303 DePaul Cr.
Bridgeton, MO 63044
(314) 344-6000

Jewish Hospital
216 S. Kingshighway
St. Louis, MO 63110
(314) 454-7000

Missouri Baptist Hospital
3015 N. Ballas Rd.
Des Peres, MO 63131
(314) 569-5200

St. Anthony's Medical Center
10010 Kennerly Rd.
St. Louis, MO 63128
(314) 525-1000

St. Elizabeth's Hospital
211 S. Third St.
Belleville, IL 62221
(618) 234-2120

St. Joseph's Health Center
300 First Capitol Dr.
St. Charles, MO 63302
(314) 947-5000

St. Luke's Hospital
232 S. Woods Mill Road
Chesterfield, MO 63017
(314) 434-1500

Hotels

St. Louis has ample luxury hotels with the addition of the Adams Mark Hotel across from the Arch on the riverfront and Omni International adjoining Union Station. These are welcome since tourism and conventions pump nearly $2 billion annually into the local economy.

Adams Mark Hotel
Fourth and Chestnut Sts.
St. Louis, MO 63102
(314) 241-7400
910-room luxury hotel next to the Old Courthouse.

The Breckenridge Frontenac
1335 S. Lindbergh Blvd.
Ladue, MO 63131
(314) 993-1100
Across from exclusive Plaza Frontenac with 310 rooms, 35 suites, seats 1100 for meetings.

Chase Park Plaza Hotel
212 N. Kingshighway
St. Louis, MO 63108
(314) 361-2500
West end hotel near Forest Park with 300 rooms and banquet space for 2,500.

Clarion Hotel
200 S. Fourth St.
St. Louis, MO 63102
(314) 241-9500
Riverfront highrise hotel, rooftop restaurant, views of river and skyline, 2000 banquet capacity and 900 rooms.

Collinsville Hilton Inn and Conference Center
1000 Eastport Plaza Dr.
Collinsville, IL 62234
(618) 345-2800
Black Swan dining room, 236 rooms.

Holiday Inn Riverfront
200 N. Fourth St.
St. Louis, MO 63102
(314) 621-8200
Dining room overlooks Missouri River, 444 rooms.

Holiday Inn West Port
1973 Craigshire Dr.
Maryland Heights, MO 63141
(314) 434-0100
Highrise hotel in suburban plaza, 325 rooms, 700 banquet seats.

Mariott Pavillion Hotel
One South Broadway
St. Louis, MO 63102
(314) 421-1776
Second largest with 672 rooms features J. W. Carver's restaurant.

Omni International Hotel
One St. Louis Station
St. Louis, MO 63102
(314) 241-6664
Located within Union Station complex, 546 rooms, American Rotisserie restaurant, plus Hart & Dierdorf restaurant nearby.

St. Louis Airport Marriott
Interstate 70 at Lambert Field
Bridgeton, MO 63134
(314) 423-9700
Directly across from Lambert Field, 602 rooms, The Hangar restaurant.

Sheraton St. Louis Hotel
910 N. Seventh St.
St. Louis, MO 63101
(314) 231-5100
Downtown hotel near shopping, 614 rooms, Plaza 900 restaurant.

Stouffer Concourse Hotel
9801 Natural Bridge Rd.
Bridgeton, MO 63134
(314) 429-1100
Mobile 4-star hotel near Lambert International Airport, Tivoli rooftop restaurant, 400 rooms.

Human Service Agencies

Excellent support exists in St. Louis for families and individuals experiencing the stress of job-hunting and joblessness. A call to the United Way office can help you pinpoint the group most likely to meet your needs. If you would like to find a job working in one of these agencies, check your professional associations and the United Way for job leads. Universities in the area may also be of assistance, particularly if you are an alum.

Catholic Family Services
4140 Lindell
St. Louis, MO 63108
(314) 371-4980 ext. 211/212

Branch Offices:
8039 Watson
Webster Groves, MO 63119
(314) 968-8010

15 St. Anthony Ln
Florissant, MO 63033
(314) 831-1533

1360 S. 5th St.
St. Charles, MO 63301
(314) 946-6014

Catholic Charities
4532 Lindell
St. Louis, MO 63108
(314) 367-5500

Family & Children's Services of Greater St. Louis
2650 Olive
St. Louis, MO 63103
(314) 371-6500

West Office:
107 South Meramec
Clayton, MO 63105
(314) 727-3235

Jewish Family & Children Service
9385 Olive
St. Louis, MO 63132
(314) 993-1000
West Office:
106 Four Seasons Center
Suite 101A
Chesterfield, MO 63017
(314) 469-3555

Lutheran Family & Children Services of Missouri
4625 Lindell
St. Louis, MO 63108
(314) 361-2121

Salvation Army (Family Service Dept.)
3744 Lindell
St. Louis, MO 63108
(314) 534-1250

United Way Community Services Directory
915 Olive
St. Louis, MO 63101
(314) 421-0700

For further information, consult the Survival Resources section.

Insurance Agencies

More than a dozen local firms trace their beginnings to the 19th century when St. Louis' companies insured the Western United States. Today a smorgasboard of companies and protection plans are available, but seven of the largest firms today opened in the 1880's or earlier. As the population ages, insurance agents face new challenges and opportunities.

Alexander & Alexander, Inc.
120 S. Central Ave.
Clayton, MO 63105
(314) 889-9200
Founded in 1898, this large firm operates nationwide with 250 employees in St. Louis.

Cervantes - Diversified Associates Inc.
52 Maryland Plaza
St, Louis, MO 63108
(314) 361-3600

Charles L. Crane Agency
10 Stadium Plaza
St. Louis, MO 63102
(314) 241-8700
A local firm founded in 1885.

The Daniel and Henry Co.
100 N. Jefferson Ave.
St. Louis, MO 63103
(314) 421-1525

Lawton-Byrne-Bruner Insurance Agency
300 S. Broadway
St. Louis, MO 63102
(314) 621-5540
St. Louis' largest agency with 275 employees, founded in 1889.

Marsh & McLennan Inc.
515 Olive St.
St. Louis, MO 63101
(314) 444-1212

Rollins Burdick Hunter of Missouri, Inc.
100 N. Broadway
St. Louis, MO 63102
(314) 241-8010

Law Firms

You can pick a specialty and find a firm to suit you, but you will need to be flexible, particularly if you're just starting out. St. Louis firms range from Bryan, Cave, one of the larger firms in the country with 200 lawyers, to small boutique firms catering to specialized clients.

Armstrong, Teasdale, Kramer & Vaughan & Schlafly
611 Olive St.
St. Louis, MO 63101
(314) 621-5070
One of St. Louis' larger firms.

Bryan, Cave, McPheeters & McRoberts
500 N. Broadway
St. Louis, MO 63102
(314) 231-8600
Largest firm in St. Louis with nearly 200 lawyers.

Evans & Dixon
314 N. Broadway
St. Louis, MO 63102
(314) 621-7755
Specializes in insurance, real estate, workmen's comp, hospital and tax law.

Greensfelder, Hemker, Wiese, Gale & Chappelow, P.C.
10 S. Broadway
St. Louis, MO 63102
(314) 241-9090
Business and construction specialist.

Husch, Eppenberger, Donohue, Elson & Cornfeld
100 N. Broadway
St. Louis, MO 63102
(314) 421-4800
Corporate lawyers with Kansas City office.

Lewis & Rice
611 Olive St.
St. Louis, MO 63101
(314) 444-7600
Large full-service local firm.

Peper, Martin, Jensen, Maichel & Hetlage
720 Olive St.
St. Louis, MO 63101
(314) 421-3850
Large primarily corporate firm.

Thompson & Mitchell
One Mercantile Center
St. Louis, MO 63101
(314) 231-7676
Second largest in St. Louis with offices in Belleville and St. Louis.

Media: Broadcast

As financial stakes rise (the number of advertising dollars increases for each ratings point), competition increases among St. Louis' radio stations.

KMOX radio (CBS-owned), St. Louis' leading station for more than 20 years, traces its success to a news/talk, sports information format. At night, it draws a national audience thanks to KMOX's 50,000-watt clear channel station. Younger audiences listen to the album-oriented rock station KSHE-FM, which has a popular morning drive-time show. KMOV-TV, once CBS-owned as KMOX, continues CBS programming. Changing newscaster and formats make for a lively, competitive market.

KETC-TV (Channel 9)
6996 Millbrook Blvd.
University City, MO 63130
(314) 725-2460

Public television station

KEZK-FM 102.5
7711 Carondelet Ave.
Clayton, MO 63105
(314) 727-2160

Easy listening station.

KHTR-FM 103.3
One S. Memorial Dr.
St. Louis, MO 63102
(314) 621-2345

Contemporary hit radio.

KMJM-FM 108
532 DeBaliviere Ave.
St. Louis, MO 63112
(314) 361-1108

Urban contemporary.

KMOV-TV
Channel 4
One S. Memorial Dr.
St. Louis, MO 63102
(314) 621-4444

Formerly CBS-owned KMOX-TV, now CBS affiliate.

KMOX-AM 1120
One S. Memorial Dr.
St. Louis, MO 63102
(314) 621-2345

Historically No. 1 radio station with a news/talk, sports information format.

KSDK-TV
Channel 5
Television Plaza
& 1000 Market St.
St. Louis, MO 63101
(314) 421-5055

NBC affiliate.

KSD FM 93.7
10155 Corporate Square
St. Louis, MO 63132
(314) 997-5594

KSHE FM 94.7
700 Union Station
Suite 700
St. Louis, MO 63102
(314) 621-0095

A move to Union Station brings traffic and a young affluent audience to this rock station as it battles KMOX.

KTVI-TV
Channel 2
5915 Berthold Ave.
St. Louis, MO 63110
(314) 647-2222

KUSA AM 550
10156 Corporate Square
St. Louis, MO 63132
(314) 997-5594

Modern country.

KYKY FM 98
111 S. Bemiston Ave.
Clayton, MO 63105
(314) 725-9814

Adult contemporary.

KWK FM 106.5
2360 Hampton Ave.
St. Louis, MO 63139
(314) 644-1380

Contemporary hit radio.

WIL FM 92.3
300 N. Tucker Blvd.
St. Louis, MO
(314) 436-1600

Only station in the top 10 playing continuous country music.

Media: Print

During the "gilded age" after the Civil War, Joseph Pultizer came to St. Louis with very little money. He became a state legislator and co-editor of the *Westliche Post*. He helped assure St. Louis' selection as the nation's 4th largest city in 1870 (despite the fact that Chicago had more people). In 1875 the *Globe* merged with the *Democrat* as a result of the Whiskey Wars during the Grant Administration. Three years later Pulitizer combined the *Post* and the *Dispatch* to create what is now St. Louis' only surviving mass circulation newspaper. Then it was the paper of the middle/lower class and often attacked city fathers as one of the first muckraking newspapers.

In recent years the vast suburbs surrounding the city have been fertile ground for a variety of community and business newspapers. Now that there is but one daily newspaper, newcomers seeking jobs as newspaper reporters, editors and photojournalists should look first to these suburban papers to offer the experience necessary before approaching larger dailies and magazines.

Clayton Leader
7 N. Brentwood Blvd.
Clayton, MO 63105
(314) 725-0125

Commerce Magazine
10 S. Broadway
St. Louis, MO 63102
(314) 231-5555
Published by the Regional Commerce and Growth Association (Chamber of Commerce).

Commerce Publishing Co.
408 Olive St.
St. Louis, MO 63102
(314) 421-5445
Publisher of Decor, Club Management, American Agent & Broker, Life Insurance Selling *and* Mid-Continent Banker.

Jefferson County News
414 Gravois
Fenton, MO 63026
(314) 343-1122

Saddle & Bridle Magazine
375 N. Jackson Ave.
St. Louis, MO 63130
(314) 725-9115

St. Louis Bride
10063 Manchester Road, Suite 105
St. Louis, MO 63122
(314) 965-6002

St. Louis Magazine
712 N. 12th St.
St. Louis, MO 63102
(314) 231-7200
A New York-style city magazine with articles about successful St. Louis' residents.

St. Louis Business Journal
712 N. Second St.
St. Louis, MO 63102
(314) 421-6200
Packed with useful local business information each week.

St. Louis Construction News & Reviews
130 West Lockwood Ave.
St. Louis, MO 63119
(314) 961-6644

St. Louis Daily Record
St. Louis Countian
612 N. 2nd
St. Louis, MO 63188
(314) 421-1880
Local and national, business and legal information from a business perspective.

St. Louis Journalism Review
8380 Olive Blvd.
St. Louis, MO 63132
(314) 991-1699

St. Louis Suburban Newspapers Inc.
1714 Deer Track Trail
St. Louis, MO 63131
(314) 821-1100
Publishing newspapers for the South Side, West County, Jefferson County, Meramec and Fenton.

The Sporting News
1212 N. Lindbergh Blvd.
St. Louis, MO 63132
(314) 997-7111

Illinois

Alton Telegraph
111 East Broadway
Alton, IL 62002
(618) 463-2500

Belleville Journal
6 N. Church
Belleville, IL 62220
(618) 277-7000

Belleville News-Democrat and Daily Advocate
120 South Illinois St.
Belleville, IL 62222
(618) 234-1000

Clinton County Journal
P.O. Box 165
Mascoutah, IL 62258
(618) 566-7925

Metal Products

St. Louis' position on the Mississippi and Missouri rivers enables companies to transport the bulky supplies and products in this business. Changing requirements for metal products have forced manufacturers to be innovative or face being squeezed out of the market.

American Can Co.
3200 S. Kingshighway Blvd.
St. Louis, MO 63139
(314) 773-2200
National manufacturer of metal and alumnium cans.

Bristol Steel Co.
3117 S. Big Bend Blvd.
St. Louis, MO 63143
(314) 644-2200

Chromally American Corp.
120 S. Central
Clayton, MO 63105
(314) 726-9200
Division of a multi-industry corporation specializing in metal fabrication; Precoat Metals Division is Midwest's largest producer of coated steel used in commercial and farm buildings.

Cerro Copper Products Co.
P.O. Box 681
East St. Louis, MO 62202
(618) 337-6000
National manufacturer of copper tubing & copper cathodes.

Consolidated Aluminum Corp.
11960 Westline Industrial Dr.
St. Louis, MO 63146
(314) 878-6950

Cupples Products
2650 S. Hanley Rd.
St. Louis, MO 63144
(314) 781-6729
Manufactures, designs and installs aluminum and stainless steel exterior wall systems.

Granite City Steel Division
20th & State Sts.
Granite City, IL 62040
(314) 451-3456
A division of National Steel Corp., mill producing flat-rolled steel.

LaClede Steel Co.
10 Broadway
St. Louis, MO 63102
(314) 425-1400
Manufacturer of carbon and steel alloy products.

Nooter Corp.
1400 S. Third St.
St. Louis, MO 63104
(314) 621-6000
Major manufacturer of steel and alloy plate, erects tanks, a subsidiary does custom thermal spraying.

National Associations
(St. Louis-Based)

St. Louis' center-of-the-country location lures organizations that want to service constituents on both coasts simultaneously. City fathers hope to draw more associations to town by mapping out the metropolitan area's strengths.

**American Association
of Orthodontists**
460 N. Lindbergh
St. Louis, MO 63141
(314) 993-1700

**American Soybean
Association**
777 Craig Rd.
St. Louis, MO 63141
(314) 432-1600

Catholic Health Association
4455 Woodson Rd.
St. Louis, MO 63134
(314) 427-2500

**Independent Computer
Consultants Associates**
443 N. New Ballas Rd.
Suite 249
St. Louis, MO 63141
(314) 997-4633

Paper & Allied Products

The area's access to the Mississippi and Missouri rivers has provided the vast amounts of water often required by this industry. Easy access to several modes of transportation also makes St. Louis a likely location for packaging. Anheuser-Busch, one of the larger companies in town, has an extensive marketing and packaging division that might be of interest to an industry professional looking for a fast-paced environment.

**Boise-Cascade Consumer
Packaging**
13300 Interstate Dr.
Hazelwood, MO 63042
(314) 344-2200

Central States Diversified
9322 Manchester Rd.
St. Louis, MO 63119
(314) 961-4300
*Manufactures a variety of paper
containers and paper medical
supplies.*

International Paper Co.
1285 Dunn Rd.
St. Louis, MO 63138
(314) 869-2220

Jefferson Smurfit Corp.
401 Alton St.
Alton, IL 62002
(314) 463-6000

Midland Container Corp.
827 Koeln Ave.
St. Louis, MO 63111
(314) 638-0028

Tension Envelope Corp.
5001 Southwest Ave.
St. Louis, MO 63110
(314) 773-7700

Weyerhaeuser Co.
120 S. Central
Clayton, MO 63105
(314) 863-8555

Petroleum Refining and Related Industries

Unstable supply prices, limited domestic resources, and shifting demand for petroleum products makes this a complex industry. Apex, one of the largest privately held companies in America, feels these changes, as do the smaller companies, but Apex has the advantage of diversification. It has been one of the largest employers in the area.

Amoco Petroleum Additives Co.
231 S. Bemiston Ave.
Clayton, MO 63105
(314) 854-8000
Headquarters for all Amoco Additive operations, manufacturer of petrochemicals, refinery at Wood River, IL.

Apex Oil Co.
8182 Maryland Ave.
Clayton, MO 63105
(314) 889-9600
Third largest privately held company in America is among the area's largest employers.

Ethyl Petroleum Additives, Inc.
20 S. Fourth St.
St. Louis, MO 63102
(314) 421-3930

Quaker State Oil Refinery
9060 Latty Ave.
St. Louis, MO 63134
(314) 521-3900

Mississippi River Transmission Corp.
9900 Clayton Rd.
St. Louis, MO 63124
(314) 991-9900

Shell Wood River Complex
P.O. Box 262
Wood River, IL 62095
(618) 254-7371
Refines crude oil, produces petrochemicals and ships products nationwide.

Printing Companies

Since some of the largest corporations in America reside here, there are ample opportunities for printers in St. Louis. Fulfilling the needs of corporations like Anheuser-Busch, Monsanto, Ralston-Purina and McDonnell Douglas provides a stable income for several printers, but there is stiff competition for these accounts. Other smaller firms do well picking up the remaining jobs. Skilled printers can usually find a niche. Most large shops are unionized, with the majority belonging to the Graphic Communication International Union Local 505. Companies listed are the largest in the area.

Color Associates Inc.
10818 Midwest Industrial Blvd.
St. Louis, MO 63132
(314) 423-9300

Color-Art Inc.
10300 Watson Rd.
St. Louis, MO 63127
(314) 966-2000

Drug Package Inc.
901 Drug Package Ln
O'Fallon, MO 63366
1-(314)-441-3830

Keeler/Morris Printing Co.
5800 Fee Fee Rd.
Hazelwood, MO 63042
(314) 731-2600

Cliff Kelley Inc.
2850 S. Jefferson Ave.
St. Louis, MO 63118
(314) 664-0023

Mid-America Printing Co.
4356 Duncan Ave.
St. Louis, MO 63110
(314) 531-8350

Nies/Kaiser Printing Co.
5900 Berthold Ave.
St. Louis, MO 63110
(314) 647-3400

Nordman Printing Co.
4210 Chippewa
St. Louis, MO 63116
(314) 773-3000

Sayers Communication Group Inc.
9600 Manchester Rd.
St. Louis, MO 63119
(314) 968-5400

St. Louis Lithographing Co.
6880 Heege Rd.
St. Louis, MO 63123
(314) 352-1300

Public Relations Firms

Few companies of any size are presenting their corporate image without the aid of outside help. As a result, several PR firms built their reputations here and have expanded throughout Missouri and beyond. John Graham, president and chief executive of Fleishman-Hillard, sees great opportunities ahead for young people in public relations. According to Graham, prerequisites for those aspiring to PR careers are: An ability to write, an ability to verbally articulate ideas, a strong ability to analyze issues and look for problems or opportunities in those issues and develop programs to minimize the problems or maximize the opportunities.

In some cases internships or co-op positions may be open at selected firms. Expect to be asked for a portfolio of your work, if not at least your ideas.

Aaron Cushman Assoc. Inc.
7777 Bonhomme Ave.,
Suite 900
Clayton, MO 63105
(314) 725-6400
Second largest firm in Missouri

Aragon Consulting Group Inc.
120 S. Central Ave.
Suite 1122
Clayton, MO 63105
(314) 726-0746
Management, marketing consulting firm.

Drolich Assoc.
1221 S. Brentwood Blvd.
St. Louis, MO 63117
(314) 862-9090
Publishes St. Louis Metro Media Guide *and* Drohlich Report.

Edelman Inc.
515 Olive St.
Suite 1605
St. Louis, MO 63101
(314) 421-6460

Fleishman-Hillard, Inc.
200 N. Broadway
St. Louis, MO 63102
(314) 982-1700

*Third largest PR firm in U.S.; owns
Wire News Network*

**International Meeting
Management**
118 S. Seminary
Collinsville, IL 62234
(618) 345-1634

Tretter-Gorman, Inc.
711 N. 11th St.
St. Louis, MO 63101
(314) 241-7704

*Emphasis on market support;
conducts market and public
opinion research; second largest
in St. Louis.*

Real Estate

Several areas in St. Louis, particularly in the St. Charles/St. Peters and Belleville regions, are very hot, perhaps because they represent some of the few properties within economic range of young families. A return to historic houses in the city has opened new markets there as well, but many at the higher end of the scale. Rising prices have discouraged some homebuyers, but St. Louis still has housing bargains. The challenge for agents is to find ways to assist potential clients in realizing their dreams. Fluctuating interest rates will have a tremendous impact on this field.

**Edward L. Bakewell Inc.,
Realtor**
7716 Forsyth Blvd.
St. Louis, MO 63105
(314) 721-5555

Century 21 - Regional Office
1982 Concourse
St. Louis, MO 63146
(314) 567-4100

Coldwell Banker - Ira E. Berry
7711 Bonhomme Ave.
Clayton, MO 63105
(314) 725-9880

Dolan Co. Realtors
8137 Forsyth Blvd.
Clayton, MO 63105
(314) 863-2610

**Gundaker Realtors - Better
Homes & Gardens**
940 West Port Plaza
St. Louis, MO 63146
(314) 576-0600

The Henry Co. Realtors
13448 Clayton Rd.
Des Peres, MO 63131
(314) 878-9911

McKelvey Realty Co.
2230 First Capitol Dr.
St. Charles, MO 63301
(314) 946-3990

L.K. Wood Realty Co. Inc.
5600 Hampton Ave.
St. Louis, MO 63109
(314) 352-7400

Voges Co. Inc.
11327 Gravois Rd.
Sappington, MO 63126
(314) 843-6400

Retailers and Wholesalers

With retail sales topping $14 billion in the Metropolitan area, there's plenty of room for talented newcomers. Buy-outs and mergers of regional companies with larger national firms continue here as nationwide, causing some dislocations, but opening opportunities for others.

Dillard's Midwest
145 Crestwood Plaza
St. Louis, MO 63126
(314) 968-5890
*Department store chain head-
quartered in Arkansas which
expanded into St. Louis and
Kansas City in 1984 when it
purchased Stix Baer & Fuller.*

Famous Barr Co.
601 Olive St.
St. Louis 63101
(314) 444-3111
*Largest department store chain in
St. Louis; founded in 1892.*

Interco Inc.
10 Broadway
St. Louis, MO 63102
(314) 231-1100
*Leading national manufacturer
of consumer goods ranging from
clothing to footwear and furniture
including International Shoe Co.
and Converse Inc.*

May Department Stores Co.
611 Olive St.
St. Louis, MO 63101
(314) 342-6300
*Parent company of Famous Barr
Co. and one of nation's largest
department store chains following
acquisition of Associated Dry
Goods of New York City (Macy's
and Lord & Taylor).*

Neiman-Marcus
100 Plaza Frontenac
St. Louis, MO 63131
(314) 567-9811

J.C. Penny Co.
4240 Rider Trail N.
Earth City, MO 63045
(314) 739-8411

Sears Roebuck & Co.
490 Northwest Plaza
(314) 344-5600

Securities and Commodity Brokers and Dealers

The leading national brokerage firms are represented in St. Louis along with several family-run local firms with excellent reputations. Tighter restrictions on trading in stocks and bonds may be mandated as a result of events on Wall Street, but these changes could bring exciting opportunities for newcomers and seasoned pros alike.

American Capital Equities
111 West Port Plaza
St. Louis, MO 63146
(314) 878-1113

A.G. Edwards & Sons Inc.
One N. Jefferson
St. Louis, MO 63103
(314) 289-3000

Dean Witter Reynolds Inc.
111 West Port Plaza
St. Louis, MO 63146
(314) 576-4600

E.F. Hutton & Co.
100 N. Broadway
St. Louis, MO 63102
(314) 231-9580

I.M. Simon & Co.
7730 Forsyth Blvd.
Clayton, MO 63105
(314) 862-8800

Newhard Cook & Co. Inc.
300 N. Broadway
St. Louis, MO 63102
(314) 342-4000

Merrill Lynch, Pierce, Fenner & Smith Inc.
1010 Market St.
St. Louis, MO 63101
(314) 982-8000

R. Rowland & Co.
100 N. Broadway
St. Louis, MO 63102
(314) 342-2800

Shearson Lehman Brothers Inc.
One Mercantile Center
St. Louis, MO 63101
(314) 444-5000

Stifel Financial Corp.
500 N. Broadway
St. Louis, MO 63102
(314) 342-2000

Shoes and Textiles

"First in shoes, booze and the American League" describes St. Louis' reputation for much of its history. Today offshore competition has forced changes in the American market, but St. Louis' Brown Group is still the nation's largest shoe manufacturer and one of the largest shoe retailers. Another local company, International Shoe, purchased Converse Inc., the athletic footwear company.

Belleville Shoe Manufacturing Co.
100 Premier Dr.
Belleville, MO 62220
(618) 233-5600

Biltwell Co. Inc.
2005 Walton Rd.
St. Louis, MO 63114
(314) 426-3850
Manufactures John Alexander men's clothing and contemporary women's clothing; subsidiary of Interco.

Bridal Originals
1717 Olive St.
St. Louis, MO 63103
(314) 436-0070

The Brown Group Inc.
8400 Maryland Ave.
Clayton, Mo 63105
(314) 854-4000
Largest American shoe manufacturer with 20 plants in operation; closed 12 plants due to changes in American shoe buying.

Edison Brothers Shoes
501 N. Broadway
St. Louis, MO 63102
(314) 331-6000
Leading St. Louis company with offices next to St. Louis Centre.

Hazel Co.
1200 S. Stafford
Washington, MO 63090
(314) 239-2781
Manufactures leather and vinyl briefcases, record binders; purchased by Josten.

International Shoe Machine Co.
4352-A Rider Trail N.
Earth City, MO 63045
(3140 739-6473

Kangaroos USA
1809 Clarkson Rd.
Chesterfield, MO 63017
(314) 532-3361

Kellwood Co.
600 Kellwood Parkway
P.O. Box 14374
St. Louis, MO 63178
(314) 576-3100
Manufactures of Sears clothing.

Monsanto Corp.
800 N. Lindbergh Blvd.
St. Louis, MO 63167
(314) 694-1000

National Garmet Co.
4284 Rider Trail N.
Earth City, MO 63045
(314) 291-8540
Makes children's sportswear.

Wolff Shoe Co.
1705 Larkin Williams Rd.
Fenton, MO 63026
(314) 343-7770
Imports, manufactures and sells wholesale women's high fashion shoes; founded in St. Louis in 1918.

Steel Products

Steel products coming in from Japan and South Korea are eating into the American market, depressing the local economy. Some types of specialty steel are useful to local industry. Investigate the companies fully before you sign on the dotted line.

Bristol Steel Co.
3117 S. Big Bend Blvd.
St. Louis, MO 63143
(314) 644-2200

Cupples Products
2650 S. Hanley Rd.
St. Louis, MO 63144
(314) 781-6729
Designs, manufactures and installs aluminum and stainless steel walls.

Granite City Steel Division
20th & State Sts,
Granite City, IL 62040
(618) 451-3456
Steel mill producing flat-rolled steel; division of National Steel Corp.

Nooter Corp.
1400 S. 3rd St.
St. Louis, MO 63104
(314) 621-6000
Major manufacturer of steel and alloy plate.

Olin Corp.
Shamrock St.
East Alton, IL 62024
(314) 258-2000

Transportation and Shipping Companies

St. Louis' location near the center of the country guarantees a ready market of people and products on the move. TWA located its headquarters here for exactly that reason, as did Ozark Airlines, now merged into TWA. Long before the airlines, St. Louis and the Midwest depended on the Missouri and Mississippi rivers for travel. Today barges ferry Midwestern agricultural products, petrochemicals, and oil and gas from silos and refineries along the route.

Agri-Trans Corp.
10825 Watson Rd.
St. Louis, MO 63127
(314) 965-4700

American Transit Corp.
120 S.Central Ave.
Clayton, MO 63105
(314) 726-9200
Bus transit division of Chromalloy.

CLC of America, Inc.
1655 Des Peres Rd.
P.O. Box 31090
St. Louis, MO 63131
(314) 966-3757
River transportation and trucking.

Eastern Airlines
Administration Office
Lambert Field-St. Louis
International Airport
P.O. Box 10175
St. Louis, MO 63145
(314) 427-5440

Midcoast Aviation
Lambert Field-St. Louis
Box 10056
St. Louis, MO 63145
International Airport
St. Louis, MO 63145
(314) 426-7060

Trans World Airlines
Regional Hdq.
500 Northwest Plaza
St. Ann, MO 63074
(314) 344-3428

Travel Services

One of the fastest growing careers nationwide, travel and tourism is a particularly important aspect of the local economy. An increase in convention business in St. Louis has opened up new jobs. The Maritz Travel Company, headquartered here, handles much of the business travel for many St. Louis-based companies. It is also one of the larger travel companies in the nation and has branched out into other related businesses.

Apex Travel
1010 Market St.
St. Louis, MO 63101
(314) 621-1010

Ask Mr. Foster
1850 Craigshire Rd.
St. Louis, MO 63146
(314) 434-1220
Parent is largest independently owned travel firm in North America.

Thomas Cook Travel
7777 Bonhomme
Clayton, MO 63105
(314) 727-3000

First Travel Co.
1399 Manchester Rd.
St. Louis, MO 63122
(314) 227-4006

Gwin's Travel Planners
111 N. Taylor Ave.
Kirkwood, MO 63133
(314) 822-1940

INTRAV Inc.
7711 Bonhomme Ave.
Clayton, MO 63105
(314) 727-0500
Affiliated with Clipper Cruise Line.

Let's Travel Inc.
450 N. Lindbergh Blvd.
Suite 101
Creve Coeur, MO 63141
(314) 991-2860

London International Travel Ltd.
11505 Olive Blvd.
Creve Coeur, MO 63141
(314) 567-6577

Maritz Travel Co.
1385 N. Highway Dr.
Fenton, MO 63026
(314) 827-4000

Service Travel Co and Horizon Tours
35 Northland Center
St. Louis, MO 63136
(314) 389-7777

The Travel Co.
11457 Olive St. Rd.
Creve Couer, MO 63141
(314) 432-6020

Utilities

The break-up of Southwestern Bell has forced the company to dismiss employees in St. Louis and throughout the Bell system. Future technological developments in the delivery of other utilities, like electricity and gas, may force layoffs of people involved with installation of out-dated equipment or services.

But these developments will also open alternative employment for the people who are trained and prepared for the future. Keeping up with professional literature and attending pertinent conferences will help employees be a step ahead of the changes.

Laclede Gas Co.
720 Olive St.
St. Louis, MO 63101
(314) 342-0500

Southwestern Bell
1010 Pine St.
St. Louis, MO 63101
(314) 247-9800

Union Electric Co.
1841 Gratiot
St. Louis, MO 63166
(314) 621-3222

WORKING FOR THE GOVERNMENT

Budget restraints are limiting the number of jobs available in the public sector, but there are still openings available. If working for the government interests you, the following list includes the numbers and addresses for Federal, State and County job information centers.

Federal Job Information Center
Office of Personnel Management
Old Post Office
815 Olive St.
St. Louis, MO 63101
(314) 425-4285

Phone information is available from 8 to 11 a.m. Monday, Wednesday and Friday. You can obtain self-service, walk-in information during the same hours.

Missouri Division of Employment Security
421 E. Dunklin
P. O. Box 59
Jefferson City, MO 65104
(314) 751-3215

Job vacancies listed at any local Employment Security Office.

Illinois Dept. of Employment Security
4519 W. Main St.
Belleville, IL 62223
(618) 233-4735

Regional office with full job/employment search capabilities.

St. Louis Dept. of Personnel
City Hall, Rm 100
St. Louis, MO 63103
(314) 622-4308

St. Louis County Div. of Personnel
41 S. Central
Clayton, MO 63105
(314) 889-2427

Vacancies are posted outside the office and volumes of job titles are available.

WORKING FOR YOURSELF

Starting your own business requires even more advance preparation, planning, research and networking than the job hunting process. Listed below are several organizations to help you make a success of your venture. Some offer financial information, technical training and/or management assistance for new business owners. When you are developing plans for your business, don't neglect to look into Small Business Administration workshops and management courses at the community colleges and universities. If working for yourself is your goal, take advantage of the services listed below and get to know other entrepreneurs to form your own support network.

Fred J. Borgers
553 Happy Court
Ballwin, MO 63011
(314) 227-6716

Consulting service in entrepreneurship and business management, a member of the American Society for Training and Development and former Monsanto employee.

Business Assistance Center
City Hall, Rm 421
St. Louis, MO 63103
(314) 622-4120

Liaison between the city government and the business community.

St. Louis County Economic Council
130 S. Bemiston
Suite 800
Clayton, MO 63105
(314) 721-0900

Parent organization for five Economic Development Agencies, established to assist industrial and business development in the metropolitan area.

St. Louis Regional Commerce and Growth Association
100 S. 4th St., Suite 500
St. Louis, MO 63102
(314) 231-5555

Promotes business development in a 10-county region and acts as St. Louis' Chamber of Commerce.

Leadership Council of Southwestern Illinois Southern Illinois University
P.O. Box 1029
Edwardsville, IL 62026
(618) 692-2156

Group of private and public leaders of business, industry, labor and education in Madison and St. Clair counties that encourages business investment.

Missouri Department of Economic Development
301 West High Street
Jefferson City, MO 65101
(314) 751-4982

*Small business assistance office offering a Missouri Loan Guarantee up to 90 percent guaranteed for start-up or expansion of an existing business or MO BUCKS, providing working capital for business or agriculture. Assist-*ance in developing a marketing/ business plan and obtaining the. required licenses for starting a new business.*

National Association of Women Business Owners
906 Olive St.
Penthouse
St. Louis, MO 63101
(314) 367-5300

An umbrella organization for women business owners, retailers, distributors, contractors and manufacturers.

Small Business Administration
815 Olive St.
Room 242
St. Louis, MO 63101
(314) 425-6600

SCORE (Service Corps of Retired Executives) offers counsel from 9:30 a.m. to 2:30 p.m. daily on any topic an entrepreneur might select, including, but not limited to, accounting, increasing sales, promotions, plant management, purchasing and collections. SBA will mail you information about its loan program, an application for SCORE assistance, and a list of business development brochures. SCORE offers a one-day seminar on starting and managing a new business, call for details.

NETWORKS AND CIVIC ORGANIZATIONS

Developing a personal network of contacts is one of the best ways to find a job. Consider joining a professional organization in your field of interest or one of the many networking groups in the area. Due to space limitations not all organizations have been listed.

Business and Professional Organizations

If you've been a member of a professional organization in another city or earlier in your career, now's the time to transfer or update your membership. The opportunity to meet and exchange ideas with your peers and interact in a non-stress environment far outweighs the initial financial bite of membership. Many organizations publish newsletters with helpful business tips, some provide job placement services and most hold regular meetings, where you can make invaluable contacts. In some instances, it is not necessary to join in order to attend meetings.

For further listings, check the *Sorkins Directory* at your local library or read the *St. Louis Business Journal*. The *St. Louis Post Dispatch* also lists professional meetings in the Business section each Monday.

American Institute of Architects
911 Washington St.
St. Louis, MO 63101-1203
(314) 621-3484

Monthly walking tour second Sunday of the month, educational workshops. AIA's library is open from 9 a.m.-3 p.m. M-F.

Advertising Federation of St. Louis
10877 Watson Rd.
Suite 100
St. Louis, MO 63121
(314) 966-8670

Sponsors the Flair/Addy awards for excellence in advertising and public relations.

Advertising Club of Greater St. Louis
440 Mansion House Center
St. Louis, MO 63102
(314) 231-4185

American Association of Industrial Management
8514 Eager Rd.
St. Louis, MO 63144
(314) 968-3600

Communications roundtables in 31 areas of industrial management. Plant managers and individual groups meet monthly to discuss topics such as productivity and personnel management. Research library available for business research 8 a.m. - 5 p.m. weekdays.

American Institute of Banking
818 Olive St., Suite 518
St. Louis, MO 63101
(314) 241-9280

Meets every third Thursday evening and sponsors classes and seminars in banking leading to an associate degree in banking and finance (AIB diploma).

American Society for Training and Development
13035 Olive St. Rd.
Suite 119
St. Louis, MO 63141
(314) 576-9119

Professional organization for trainers and consultants in management, organizational development and human resource planning.

Association of General Contractors of St. Louis
2301 Hampton Ave.
St. Louis, MO 63139
(314) 781-2356

Meets monthly either the first or second Thursday night for dinner and a program. Contractors and representatives of the six basic trades are eligible for membership. Suppliers may belong as affiliated members.

Association of Young Printing Executives
321 N. Spring St.
St. Louis, MO 63108
(314) 531-1610

Members of the Printing Industries of St. Louis, selection based on sales volume of executive's company, meets first Tuesday of the month.

The Bar Association of Metropolitan St. Louis
One Mercantile Center, Suite 3600
St. Louis, MO 63101
(314) 421-4134

St. Louis County:
7777 Bonhomme, 23rd Floor
Clayton, MO 63105
(314) 721-6422

Open to all attorneys. Legal education programs and a job listing book available to members.

Engineers Club of St. Louis
4359 Lindell Blvd.
St. Louis, MO 63108
(314) 533-9333

Umbrella organization for 53 engineering organizations in the Greater St. Louis area. Sponsors educational seminars and social functions for members. Student and junior memberships availa-

ble. Call for information about specialized engineering organizations (i.e. chemical and civil engineers)and meeting times.

Federal Business Association
210 N. Tucker, Room 1111
St. Louis, MO 63101
(314) 425-5715
Contact: Virginia Weidle

All federal employees are eligible to belong. Meets the second Tuesday of the month at Garavelli's Restaurant, 4630 Lindell Blvd.

International Association of Business Communicators
13035 Olive St. Rd.
Suite 119
St. Louis, MO 63141
(314) 436-7600
Contact: Susan Ruland

Publishes a job listing guide that includes educational program information.

International Association for Financial Planning
440 Selma Ave.
St. Louis, MO 63119
(314) 962-1225

An umbrella organization for financial bankers, CPAs, and lawyers working in financial planning. Application process required for membership. Educational program/dinner meetings the third Wednesday of the month, usually at Schneithorst's Hofamberg Inn (Lindbergh & Clayton).

Hospital Association of Metropolitan St. Louis
720 Olive St.
St. Louis, MO 63101
(314) 421-3415

Represents the 48 hospitals in the Metropolitan area and has information about health care careers.

Independent Computer Consultants Associates
933 Gardenview Office Pkwy.
St. Louis, MO 63141
(314) 997-4633

National Association of Accountants
Touche, Ross & Co.
2100 Railway Exchange
St. Louis, MO 63101

Meetings are held the third Tuesday of the month from 5:30-9:30 p.m. Students are eligible for membership.

St. Louis Association of Life Underwriters
818 Olive St. Suite 974
St. Louis, MO 63101
(314) 241-1418

Large organization of persons who have completed or are under training to be Life Underwriters. Meets usually the latter part of the month for an educational program.

St. Louis Metropolitan Medical Society
3839 Lindell Blvd.
St. Louis, MO 63108
(314) 371-5225

Sales & Marketing Executives of Metropolitan St. Louis
200 S. Hanley Rd.
Suite 502
Clayton, MO 63105
(314) 727-0524

Organization of executives who have the primary sales or marketing responsibilities within their companies. Meets the first Friday of the month for noon luncheon at the Holiday Inn in Clayton.

Women's Business and Professional Organizations

Some of these organizations began several years ago when women were not welcome in professional organizations then dominated by men. Times have changed, but many of the organizations continue and serve as excellent networks for area professionals. Other organizations have surfaced more recently, again offering new opportunities for upwardly mobile women.

American Institute of Banking Associates
818 Olive St.
Paul Brown Building, Suite 518
St. Louis, MO 63101
(314) 241-9280

Women's section of the educational branch of the American Banker's Association.

American Medical Women's Association
224 S. Woods Mill Rd.
Chesterfield, MO 63017
(314) 878-2556
Contact: Dolores Tucker

American Society of Women Accountants
427 Greeley Ave.
Webster Groves, MO 63119
(314) 961-2026
Contact: Sue Wolverson

Requires a minimum of two years work experience in the field.

Students with less than two years experience are accepted as junior members.

Business and Professional Women

Missouri Federation of Business and Professional Women's Clubs, Inc.
40 Plaza Square, Apt. 101
St. Louis, MO 63103
(314) 436-0939

Downtown organization of working women to promote professional opportunities through education and cooperation.

Commercial Real Estate Women of St. Louis (CREW)
7730 Forsyth Blvd.
Clayton, MO 63105
(314) 863-1500
Contace: Jane Carlos

Monthly educational programs for women actively engaged in commercial/industrial real estate brokerage and/or leasing.

CWI Credit Professionals
6500 Chippewa
Suite 225
St. Louis, MO 63109
(314) 752-9535

Educational association for people in the credit industry.

Executive Women International
5500 W. Park Ave.
St. Louis, MO 63110
(314) 647-6000
Contact: Nadine Gremmler

Non-competing firms represented by individuals in executive, key administrative or executive secretarial positions working to accomplish mutual business and professional goals.

The Fashion Group of St. Louis
75 Plaza Frontenac
Frontenac, MO 63131
(314) 432-8590

Executives or administrators engaged in the creation, production, distribution or merchandising of fashions/fashion products with three years experience.

Greater St. Louis Chapter/ American Business Women's Association
P.O. Box 69157
St. Louis, MO 63169
(314) 235-3666

Must be employed at time of application.

Missouri Nurses' Association
915 Olive St.
Suite 920
St. Louis, MO 63101
(314) 421-3450

Must be a registered nurse.

Missouri Press Women
801 Hanley Rd.
St. Louis, MO 63130
(314) 721-7413

Must be a professional communicator with at least one year's experience (associate membership available for students). Meets every 4th Thursday.

National Association of Women Business Owners
906 Olive St.
St. Louis, MO 63101
(314) 367-5300
Contact: Erin Shocklee

Provides encouragement and support for women business owners.

Personnel Association of Greater St. Louis
Hickey School
6710 Clayton Rd.
St. Louis, MO 63117
(314) 644-2866

For persons actively engaged in personnel administration or related fields for six months. Associate memberships available.

Professional Saleswomen of St. Louis
P. O. Box 50104
Clayton, MO 63105
(314) 889-1933

Provides opportunities to increase sales productivity through affiliation and reinforced confidence.

Professional Secretaries International
P.O. Box 16805
Clayton, MO 63105
(314) 961-8889

Strives to promote and enhance the secretarial profession through educational information.

St. Louis Society of Women Certified Public Accountants
P.O. Box 11844
Clayton, MO 63105
(314) 434-0822

Must be a CPA or a professional at a public accounting firm, provides career services.

St. Louis Women's Caucus for Art
P.O. Box 24181
St. Louis, MO 63130
(314) 391-9691

Educational and service organization representing and working to advance the concerns of women artists, art historians, museum professionals and other visual arts professionals.

Society of Women Engineers
c/o Engineer's Club
of St. Louis
4359 Lindell Blvd.
St. Louis, MO 63108
(314) 533-9333
Need to be actively engaged in engineering or be a graduate engineer.

Women in Energy
P. O. Box 74A
St. Louis, MO 63166
(314) 554-3258
Provides forums for discussion, distributes information and provides networking for energy professionals.

Women's Chamber of Commerce
3817 Russell Blvd.
St. Louis, MO 63110
(314) 772-4989
Opens avenues of information, assistance and support for the business, career and civic-minded woman.

Women's Council of Realtors
c/o Coldwell Banker, Ira Berry
7711 Bonhomme
Clayton, MO 63105
(314) 725-9880
Prepares women for leadership roles in business and community service.

Women in Communication
P.O. Box 9145
St. Louis, MO 63117
Publishes a newsletter for members, sponsors monthly luncheons and a job referral network.

Women Lawyers' Association of Greater St. Louis
P.O. Box 50401
Clayton, MO 63105
(314) 727-7122
Contact: Linda Murphy
Must be a member of the Bar, not necessarily in Missouri. Dues: $30.

Women's Networks

These organizations offer a place or a time to share ideas and work for personal and community development. Some are limited to members of a particular company but work for the benefit of women throughout the St. Louis area.

American Association of University Women
7359 Forsyth
Clayton, MO 63105
(314) 721-1961

The Lifelong Learning Connection (The Connection)
36 Four Seasons Center
P.O. Box 30
Chesterfield, MO 63017
(314) 878-8708
Enlightenment, Encouragement, Enrichment and Enjoyment by and for all women of all ages and stages of life. Non-denominational, non-political and not-for-profit.

Jobhunters' Support Group
P.O. Box 11517
Clayton, MO 63105
(314) 966-1120
Meets twice a month, every other Saturday, with guest speaker once a month. Job referrals, career counseling and resume assistance sponsored by the Women's Commerce Association.

Junior League of St. Louis
8250 Clayton Rd.
St. Louis, MO 63117
(314) 863-9058
Organization addressing community issues through the participation and leadership of trained women volunteers.

The Network — Monsanto
800 N. Lindbergh
St. Louis, MO 63167
(314) 694-4451

To provide management and professional skill development among Monsanto women.

National Organization for Women
P.O. Box 16132
Clayton, MO 63105
(314) 727-5466

St. Louis Women's Commerce Association
P.O. Box 11517
Clayton, MO 63105
(314) 256-2312

Promotes women in business and the professions through contacts within the community.

Southwestern Bell Professional Women
One Bell Center, Rm 38T7
St. Louis, MO 63101
(314) 235-3730

Promotes business contacts and career advancement for management women at Southwestern Bell.

University of Missouri — St. Louis Women's Center
8001 Natural Bridge Road
Clark Hall
St. Louis, MO 63121
(314) 553-5380

Creative programming and active referral network and library.

Women in Business Network Ralston-Purina
One Busch Place
St. Louis, MO 63118
(314) 577-2848

One of the largest women's organizations in St. Louis with 900 women employees providing professional enrichment opportunities.

Women in Business Network— Ralston-Purina
Checkerboard Square
St. Louis, MO 63164
(314) 982-3524

YWCA of Metropolitan St. Louis
1015 Locust St.
Suite 734
St. Louis, MO 63101
(314) 421-2750

Providing programs and coordinating resources to help women reach their fullest potential.

ILLINOIS

Every Woman's Center
Belleville Area College
2500 Carlyle Rd.
Belleville, IL 62221
(618) 274-4675

Offers support groups, personal and career counseling and job referrals.

SURVIVAL RESOURCES

Being unemployed or in a job transition can be a stressful time. If you have been terminated from your job, the pressures from financial insecurity and the stress of being labeled "unemployed" are tremendous. These pressures affect your attitude and self-image and are difficult to hide in your interview presentation. Living expenses continue and there can be increased strain on interpersonal and family relationships. You may need help in managing your transition. Whether it's a financial difficulty or a personal problem you need to work out, there are a variety of support services available in the Metropolitan Area. We've listed a few below. If you need assistance in a hurry, the United Way can refer you to one of more than 1,000 social services listed in its 1986 Community Service Directory.

United Way of Greater St. Louis
915 Olive
St. Louis, MO 63101
(314) 421-0700
More than 1,000 not-for-profit voluntary and tax-supported agencies and organizations operating in health, social welfare, recreation and education in the bi-state St. Louis area are noted in the Community Service Directory. Free, confidential service is available from 8:15 a.m. to 5 p.m. weekdays by calling "First Call For Help," a new information and referral service.

Dial HELP
325 N. Newstead
St. Louis, MO 63108
(314) 371-4357
General information and referral services weekdays: 8 a.m. - 8 p.m. Weekends: 10 a.m. - 6 p.m.

Mental Health Information

Mental Health Association of St. Louis
2510 S. Brentwood Blvd., Suite 212
St. Louis, MO 63144
(314) 961-5957

Crisis/Suicide Intervention
Life Crisis Services
Hotline: (314) 647-HELP

Mental Health Referrals
(314) 647-4357

Care & Counseling, Inc.
12145 Ladue
St. Louis, MO 63141
(314) 878-4340
Individual, marriage and family counseling provided by pastoral counselors in the Greater St. Louis Area. Fees based on ability to pay.

Family and Children's Services
2650 Olive
St. Louis, MO 63103
(314) 371-6500

Offices available throughout the city, call to locate the one nearest you.

Child Care

Child Day Care Association of St. Louis
915 Olive, Suite 913
St. Louis, MO 63101
(314) 241-3161

Referrals to help parents locate and select day care programs.

Consumer Credit

Consumer Credit Counseling Service
3833 Gravois
St. Louis, MO 63116
(314) 773-3660

Counsels individuals by appointment 8:30 a.m. - 4:30 p.m. weekdays.

Legal Services

Lawyer's Reference Service
First Floor Mezzanine
Civil Courts Bldg.
Tucker and Market
St. Louis, MO 63101
(314) 622-4995

County Government Center
Clayton County Courthouse
7900 Carondelet, Rm 355
Clayton, MO 63105
(314) 889-3073

Services for those who need a lawyer, $5 registration fee, other charges arranged with the lawyer.

Legal Aid Society
625 N. Euclid
St. Louis, MO 63108
(314) 367-1700

Provides legal assistance for those unable to afford their own defense.

Unemployment Insurance

Employment Security offices have information about Unemployment Insurance eligibility. For locations of claims offices in your area, look for Employment Security listing under the State Government Office in the blue pages of the telephone book.

LIBRARIES, DIRECTORIES AND PUBLICATIONS

When you are researching major industries and specific companies or organizations in the St. Louis area, use the local directories and publications listed here to help uncover product or service information. Changes in management and biographical data about executives can usually be found in the business press. Most publications are available at major libraries noted below.

Libraries

East St. Louis Public Library
405 N. Ninth St.
E. St. Louis, MO 62201
(618) 874-7280

St. Louis Public Library
1301 Olive St.
St. Louis, MO 63103
(314) 241-2288
An excellent source of business and career planning information, often overlooked by jobseekers. Librarians knowledgeable about St. Louis and mapping career strategy.

St. Louis County Library Plaza Frontenac Hdqs.
1640 S. Lindbergh Blvd.
Frontenac, MO 63131
(314) 994-3300
Large business and industry section, plus numerous profes-

sional newsletters and business publications.

Thomas Jefferson Library University of Missouri-St. Louis
8001 Natural Bridge Rd.
St. Louis, MO 63121
(314) 553-5050
University library with large business section.

Wolfner Memorial Library
2002 Missouri Blvd.
Jefferson City, MO 65101
(314) 751-8720
State government library complete with services for the blind and handicapped.

Directories

Book of St. Louis
712 N. Second St.
P. O. Box 647
St. Louis, MO 63188
(314) 421-6200
Guide to the leading local companies and professional organizations in St. Louis, published by the St. Louis Business Journal.

St. Louis Regional Commerce and Growth Association
100 S. 45h St., Suite 500
St. Louis, MO 63102
(314) 231-5555
Published an annual report giving the city's progress in a nutshell, names city leaders and areas of growth.

State Industrial Directory
State Department of Commerce
Chamber of Commerce of United States
1615 H St., N.W.
Washington, D.C. 20006
(202) 659-6000
Lists information about manufacturers their products and production locations.

Publications

Sorkins' Directory of Business and Government
744 A Spirit of St. Louis Blvd.
Chesterfield, MO 63017
(314) 537-1200

The guide to leading industries, businesses and those that run them in the bi-state area. Also lists community and educational leaders.

St. Louis Business Journal
612 N. Second St.
St. Louis, MO 63102
(314) 421-6200

An up-to-the-minute report of area business and industry.

St. Louis Countian
111 S. Bemiston, Suite 504
Clayton, MO 63105
(314) 727-6111

Reporting business and legal news since 1912, publishes Tues. - Sat.

St. Louis Daily Record
612 N. Second St.
P.O. Box 88910
St. Louis, MO 63188
(314) 421-1880

Publishing legal and business news Tues.-Sat. since 1890.

St. Louis Magazine
612 N. Second St.
St. Louis, MO 63102
(314) 231-7200

Trendy New York style magazine about the city and its people.

To: Career Management Press
8301 State Line, Suite 202
Kansas City, MO 64114

A resource not listed in your '87 St. Louis edition, that I have discovered, and that I want you to know about is

_____
_____
_____
_____
_____
_____
_____
_____
_____
_____
_____
_____
_____
_____
_____
_____
_____
_____
_____
_____
_____
_____
_____
_____
_____
_____
_____
_____

Or, please find the enclosed article which I thought you might enjoy reading.

Name _____
Address _____
City _____ State _____ Zip _____